AF761607

Master Keying Text Book

by Don O'Shall and Tony West

© 2017 in Beverly Hills, FL - USA
ALL RIGHTS RESERVED

Locksmithing Education Publishing, LLC
Third Edition, © August, 2017
(First Edition, 1984 Allentown, PA)
by Donald O'Shall and Tony West,
Beverly Hills, FL - USA

All rights reserved.
No part of this book may be reproduced
in any form or by any means, without
permission in writing from the publisher.

ISBN 978-1-937067-09-0

Library of Congress Cataloguing-in-Publication Data

O'Shall, Donald and
West, Anthony

Master Keying Text Book 2008 edition / by Don O'Shall and Tony West

ISBN 978-1-937067-09-0
1. Industry—Security Measures—Management

ISBN 9781937067090
$65.00 USD
56500 >
9 781937 067090

Locksmithing Education Publishing, LLC
12 South Lee Street
Beverly Hills, FL - 34465

Knowledge is built upon knowledge. Honor the past and you prepare for the future.

This is dedicated, like a drop of fine wine upon the ground, to those who have gone before us, without whom our progress would not be possible.

Not Shown: Cal Brittel, Bert Michaels

The Grand Masters Of Master Keying

"Before we move ahead we should take a moment to remember all the Master keying Experts who brought us to this point. The list is far too long to put here in its entirety, of course, so I will merely list some of those who directly influenced my own development. The following is excerpted from a speech I made in 1999 on the topic of Master keying.

"Master keying is a well established process. According to **Tom Hennessy**, (curator of the **Lock Museum of America** in Terryville, Connecticut since its foundation in 1972 and developer of the Standard Key Coding System in popular usage today), Master keying of Lever Locks can be traced back to as early as 1686 , and Master keying of Pin Tumbler Locks appeared in an 1865 Yale & Greenleaf advertisement.

But in spite of that, it was not until the 1950's when mainstream locksmiths first began to seriously evaluate its capabilities and limitations and the methods then in use to accomplish it. The **Locksmith Ledger** published its first master keying article using the techniques on which modern master keying is based, and serious discussion and investigation of the topic began.

Jerome Brush put out the **Ilco Master Key Selector** to lay out small systems.

And it was nearly twenty years later before innovative locksmiths (such as **Wiegend Jensen, Ken Ehrenreich, Bert Michaels, Hollis DeVines, Gerry Finch, Cal Brittel, as well as others too numerous to mention here, but whose efforts were more than notable**) began to explore new methods of creating the systems, and to explore questions pertaining to those perceived capabilities and limitations, and more importantly, to share their observations with others, both in person and in print (in magazines such as **The National Locksmith** and the **Locksmith Ledger**, as well as in books and courses designed primarily for the mainstream Locksmiths).

Inspired by these men (the above list represents some of those who directly influenced my own education and direction.) a new generation of Master keying Specialists emerged, and began to organize methods, formats and styles that maximized the usefulness of Professional Master keying Techniques to the public at large, as well as the important task of sharing that information not only with the mainstream Locksmith, but with a new generation of Institutional Locksmiths as well.

Once again, the list of people who should be included here is too numerous to mention, so I will concentrate on those with whom I had direct and personal dealings, often challenging each other to develop ever increasing levels of proficiency, in many cases answering questions that had not even been asked before.

These included **Billy B. Edwards, Anthony West, A. J. Hoffman, Jerome V. Andrews, Edward M. Greene, Breck Camp, Gale Johnson, Greg Brockway, John Truempy, Tony West, and of course, myself.**

I have every faith that our works will inspire yet another generation of research on the topic, and add new names to the list of those who influenced the professional techniques of Master keying."

- Don O'Shall, in a speech made in March 1999.

That faith has by the time of this writing, inspired two other locksmiths to take up the mantle. Eric Costley, younger brother of renowned lock and safe instructor Brian Costley, has been researching and writing on the Spherical Method of Master Keying. Rod Oden has been writing on the effect of the effect of Generics and Associated Master Keys on the Spherical Method. It is this author's fervent and constant hope that others will become inspired by these writings and the work that preceded them, and we will have a new generation of Master Keying experts to write and teach for the good of the Locksmithing profession.

- Don O'Shall, from a speech made in 2005.

And in 2007 yet another young locksmith, Dean Nickel, has begun research on what he calls "The Cubic Perspective".

- Don O'Shall, as an update comment on his 2005 speech.

Table of contents

Introduction to the Master Keying Textbook

Master Keying definitely ranks among the most profitable aspects of the Locksmithing trade.

Customers generally recognize the Master keying Locksmith as a professional. Very few other jobs can provide the Locksmith with such a level of respect as master keying can.

Master Keying is the basis of the Institutional Locksmith's stock in trade. Although Institutional Locksmiths "wear many hats" most of their duties are deeply involved with master key systems and key control in such a system.

Master Keying is (or should always be) a carefully planned system of controlled access, and provides a necessary balance between security on one side and life safety and convenience on the other. Master Key systems regulate who, within the pool of authorized users, has clear permission to enter each area at times when others do not (such as when the area is locked). They permit emergency personnel (such as Campus Police on a University) to enter areas in a crisis situation where every minute counts, without the need to sort through hundreds or thousands of keys.

Master Key systems exist because they meet the needs of the customer's facility. They are in use in virtually every major institution in this country, and their standards are well-developed and well promoted at every level of the security hardware and lock industries.

Master Key Systems are intended to put reasonable restraints upon the legitimate occupants of the building, while permitting them to carry out all of their necessary travels and duties within the structure(s) of the organization.

With very few exceptions it is NOT intended to "keep the bad guys out", although it may do so. There are other techniques and standards to perform that function, although a master key system can be an important component of such a security system, especially when higher security keys are chosen for the system.

Master Key systems may not be appropriate for every application, particularly unless higher security keyways are employed. Master Key systems have inherent weaknesses that permit a person with the right combination of knowledge, skills, tools and inventory to potentially compromise them. These risks may be managed or reduced to a reasonable level of risk, but they cannot generally be eliminated entirely. Some applications, on the other hand, increase the risks.

Apartment complexes, for example, are very vulnerable to attack. They offer unlimited individual access to anyone for merely the cost of renting an apartment. The new renter can then compromise the system by any of about a half dozen methods and create a master key allowing them to violate any other apartment in the complex. Many such complexes use lower quality locks and key systems, making this even easier. Yet the only reasons a master key is created for this application are to provide maintenance personnel access and to provide rental sales people access. There are better solutions than master key systems on residential grade locks and keys available for situations such as this!

States with licensing requirements have increased interest among commercial locksmiths by requiring continuous training, as have many Locksmith Associations. By requiring "retakes" of courses cleverly disguised as "continuing education", they hope to make it easier for each individual to become the best they can be.

They have recognized the fact that subjects such as master keying are one of the most precise and demanding of skills of the Locksmith, and they need to give each individual an opportunity to take it and to re-take it to maintain currency of knowledge and to be updated on new or improved methods without risking losing their status as an "expert".

Regardless of whether you are new to master keying or an established master keying locksmith, it is very likely that you will find this book very useful. It has been the basis of several courses and a spin-off for several other books. Nevertheless, some individuals may find other books or courses more suited to their level or their style of learning.

Lesson One: SOME BASIC TERMS

The first step in becoming a master keying professional is to acquire a thorough understanding of the various terms and symbols used in master keying. Many of these will be terms that you are already familiar with from your everyday locksmithing, while others may either be entirely different terms, or may be a new meaning for a familiar term.

The first source of terminology should always be the LIST Council Glossary of terms, which is available through many sources today, or can be viewed on-line at:

http://www.thelockman.com/gloscopy.htm

or at:

http://www.locksoft.com/gloscopy.htm

Originally started in 1982 as a project prepared for ALOA (Associated Locksmiths Of America), this has now become the industry standard used at every level. It can be viewed online at the above URL's or found in most Association links or Wholesaler catalogues, but the latest version (and most complete version) will be at the above locations.

The definitions used herein will conform to the LIST Council definitions in most applicable cases, but the LIST definitions tend to be clear and concise, which is not always the best way to define a term in a learning environment, so many times terms will be defined here in ways that promote the flow of learning in the book instead. Written tests tend to use the LIST Council Dictionary definitions so we will also show those in the glossaries of terms.

Primary Terms will be in **BOLD** print in upper case letters, while secondary terms will be in **Bold** print in mixed case letters. Major terms requiring extensive explanation will be centered on the page, as is "Master Keying" which immediately follows this. Very important points will be printed in UPPER CASE but not in bold print.

MASTER KEYING

When it is properly set up, Master Keying is a CAREFULLY PLANNED SYSTEM OF CONTROLLED ACCESS. The goal of master keying is to create a system that PERMITS each person to go wherever they need to in the building or area with a minimum of difficulty or inconvenience, yet at the same time LIMITS who can go where, allowing ONLY AUTHORIZED PERSONS to enter any given area of the system.

By using proper master keying techniques, we can permit the individual to quick and easy access to all authorized areas, perhaps with only a single key. Yet all the areas are locked to those who haven't been issued the proper key for them (except for those who have somehow acquired an illegitimate key for it by theft or discovery or by unauthorized duplication).

Master key systems are in use in virtually every hotel and motel, school and university, hospital, apartment or condominium complex, government area and office building in this country, as well as in many chain store operations. **They are in use in so many public institutions because they satisfy three of the basic needs of the institution—SECURITY, LIFE SAFETY and CONVENIENCE.**

At first glance these three may seem to be at cross-purposes with each other, but the simple fact is that security without the Life Safety and Convenience is virtually useless. For example, one way to theoretically gain security would be to have each door or area keyed so that a separate key operates it. However, a person needing access to several areas would have to carry a key for each. If a key were lost or stolen, it might not be discovered for some time. In a critical situation where time was vital, this person would have to sort through the keys for the proper one.

Before long, it is likely that this inconvenience will outweigh the individual's security awareness, and the area will either be left unlocked of the doors blocked open. Now convenience is in control of the situation, and security takes a back seat. But convenience without security will soon leave little or nothing of value remaining to be protected.

Life Safety, on the other hand, requires that certain types of doors be positive latching, which they are not when the opening is blocked open. It also requires easy access by those with a legitimate life safety need to enter the area.

Security will never be in absolute agreement with Life Safety and Convenience, but in a typical system it should balance them out when properly applied. True security comes not from one level of security but from different security layers requiring different combinations of knowledge, skills, tools and inventory. A properly applied and managed Master Key system is one effective layer. Adding keys which are restricted by law or high security adds another layer of security to the mix, but it really relies heavily on the Master Key system for its effectiveness.

A SAMPLE MASTER KEY SYSTEM TO SEE RELATED TERMS

Let's look at a simple example of a master key system in operation. We'll call our fictitious example the "COST-MOORE MOTEL". The Cost-Moore Motel consists of four buildings, plus a smaller building in front that houses a lobby, registration desk, Motel offices, and a cocktail lounge, and has a linen storage area in the rear of it.

Each of the other buildings consists of four floors of guest rooms (20 on each floor). The guest rooms are all rented on a daily basis.
When a guest checks in, he or she (or they) receive a guest room key that will open only one room in the entire Motel, although if there are a front and back door

on the room (such as those with a front door on the poolside and a rear door facing the sunset, for which view the Motel is famous), that guest's key would open either door to the room.

In Master keying terms, that guest's key is called a **CHANGE KEY** (possibly because it changes from room to room?). The two locks (front and rear) are referred to as **KEYED ALIKE**.

It was once common to stamp the keys with the **BITTING** (the combination of the cuts on the key), but this practice has generally been eliminated from most master key systems developed since the 1980's .

Each morning, 16 maids are expected to clean all the guest rooms. Each maid is assigned a floor to clean the rooms on as her area of responsibility. Rather than carrying 20 keys each (one for each room on her floor), the maids carry a key that opens all the rooms on her floor, but will not open any of the guest rooms on other floors. Each of these keys is called a **MASTER KEY**. (The Motel employees probably call it a "**Floor Master**" to show its level of authority.)

As each room in a given building has been cleaned by the maid assigned to it, it is inspected by one of the Maid Supervisors. Each Maid Supervisor inspects the rooms on the four floors of her assigned building, but has nothing to do with the floors of the other buildings. For that reason, each has a key what will open only the guest rooms in the building assigned to them. These keys would be called **GRAND MASTER KEYS**, since they control more than one **MASTER KEY GROUP** (as each floor could be called, because one master key operates all the locks on it).

The employees probably call these "**Building Masters**" since they operate all the locks in a particular building. **BUT REMEMBER that the "Building Masters" will not always be Grand Master Keys.** Building employees assign terms such as these to show the key's position in the institution, whereas terms such as Master, Grand Master, etc. are more universal.

The Head of Housekeeping has the final responsibility of making sure that not only do the maids do their work properly, but also that the Maid's Supervisors are properly checking them. For that reason, the Head of Housekeeping has a key that operates all the locks on all the guest rooms in all four buildings. Because it is controlling more than one Grand Master Key Group (i.e., building in this example), it is referred to as a **GREAT GRAND MASTER KEY**.

Incidentally, whenever a guest is in one of the rooms, and does not wish to be disturbed by the maid, etc, they push in a button on the inside (room side) of the lock, which then blocks out all keys, including the Master keys, Grand Master keys, and the Great Grand Master key. This is referred to as a **HOTEL FUNCTION LOCKSET**.

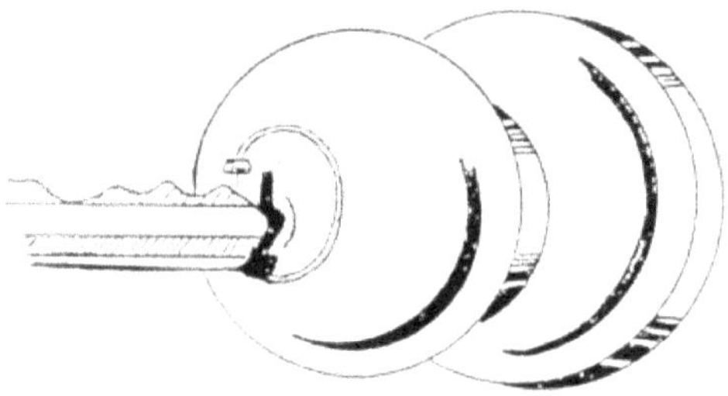

In an emergency situation, however, an **EMERGENCY KEY (or EMERGENCY MASTER KEY)** which bypasses this function is available to the Hotel Manager.

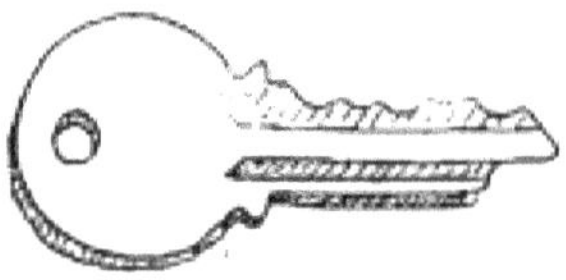

Some rooms also have available to the Hotel Manager a special key that bypasses the Hotel function on one specific room's lock(s) only. This key is used for guests such as traveling furriers who carry valuable samples with them. It is called a **SHUT-OUT KEY** or a **DISPLAY KEY**. The only difference between a Shut-Out Key and an Emergency Key is that the Emergency Key is never given to guests, since it bypasses the Hotel function on ALL the guest rooms, while the Display key only operates the individual room, locking out the maids and previous guests.

A **SPANNER KEY** (also known informally as a "snake eyes" because the two prongs on it fit two holes that resemble the two on a pair of dice, also known as "snake-eyes") is typically used to set the lock so that even when the room is temporarily unoccupied, the push button remains in the locked position, blocking all regular room or master keys.

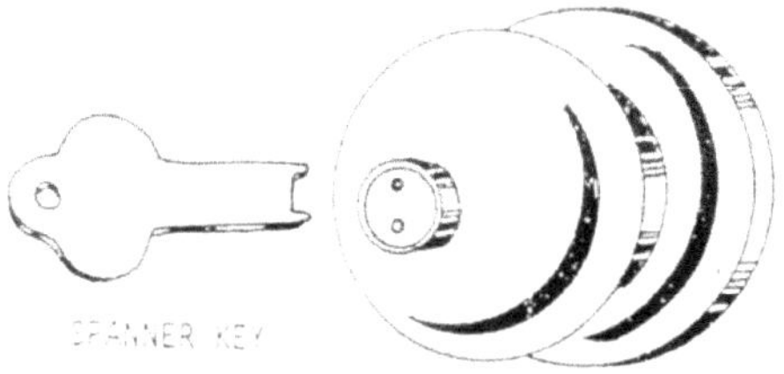

In the rear of the front building (that houses the offices, etc.) is a linen room where the maids each draw clean linens and towels for their rooms as needed. For that reason, each of the maid's Master keys, as well as the Grand Master keys used by their supervisors, and the Great Grand Master key of the Head of Housekeeping, will all open that lock. Once a day, a truck from the local laundry picks up the dirty linens and replaces them with clean. This has to be done in the presence of the Head of Housekeeping, so no individual key is issued to the laundry people. This lock would be called, in Master keying terms, **CROSS-KEYED WITHOUT ITS OWN KEY**.

If an individual key to that lock only, HAD been issued to the outside laundry people, it would have changed the term for that lock to **CROSS-KEYED WITH ITS OWN KEY**.

The term **CROSS-KEYED** is used to show that keys which normally would not operate each other's locks are able to operate that lock. Cross-keying is one of the rare instances where a term describes the lock, and not necessarily the key for it.

Other types of Cross-Keying that we find in other buildings are easy to identify. Some buildings have each tenant's key operating the front door lock. This would be a type of Cross-Keying. (It is called **Maison Keying**, in this application. **Maisson Keying** is an archaic spelling that appeared in some books. Modern French uses the single s rather than the double s that once indicated a Mansion rather than a house.)

Sometimes a secretarial pool must have access to a stockroom to draw supplies, but the stock person is not permitted into the secretarial offices. In this case, the stockroom would once again be Cross-Keyed With Its Own Key. These are only a few examples of Cross-Keying.

If this were the full extent of the Master key system, we would call it a **GREAT GRAND MASTER KEY SYSTEM,** since that is the highest level of Master Key we have seen so far. Similarly, if there were one less level (if only one Maid Supervisor existed, for example, and no one checked her work, and she checked all the rooms in all the buildings) the system would be called a **GRAND MASTER KEY SYSTEM,** since that would be the highest master key level in the system. And finally, if there were no higher Master Key in the system than the Maid's Master Key, this would be called a **SIMPLE MASTER KEY SYSTEM.**

However, we still have one building to key in the Cost-Moore, so it is likely that the system will go higher yet. None of the existing keys has any control over any of the rooms or areas in the front building. So let's look at its specific master keying needs.

The Lobby doors are always open and unlocked, so we don't need to be overly concerned with them at this time, although they should be a part of the system, just in case needs change in the future.

Behind the reception desk is an office for use by the people at the desk.

The reception office has two doors leading to opposite ends of the desk, and one door on the side that the clerks use to get to the corridor immediately behind the desk.

Because the same people use all three doors, and since they all lead into essentially the same area, it is common practice to make one key fit all three locks. This is often called **KEYED ALIKE**. That key will also be considered a **CHANGE KEY,** since the Assistant Manager and Manager both have a key that will operate these three locks, as well as others in the building.

In fact, the Assistant Manager's key operates ALL the locks in the front building, and the Manager's key operates all the locks in the entire Motel! We'll come back to that later, though.

The individual key is the same for that group of three locks which are KEYED ALIKE because it opens locks that EACH HAVE THE SAME PINNING COMBINATION in them.

A Master Key, on the other hand, operates locks that each have their OWN pinning combinations, and which are NOT all exactly the same.

The cocktail Lounge has its own key for the front door to the Lounge. That key would also be considered a change key.

The bar manager, however, also carries a key to the liquor storage room. He is the only person in the building with a copy of that key, and NONE of the Master Keys of any level (Master, Grand Master, Great Grand Master, etc.) will open that lock. We call that lock **SINGLE KEYED**. (It has also been referred to as Single Keyed Different.)

Recently, a new term has also been added that is related to the single keyed term. Sometimes the key that is intended for a single keyed cylinder, will also operate something in the system that falls under a master key of some level. In this case, the key becomes a change key, because it falls under the master key, but the lock which would normally be single keyed becomes pinned **NO MASTER KEY (NMK)** rather than simply being considered single keyed.

The Maintenance Engineer has a key that allows him to enter the Reception desk area and the Cocktail Lounge to repair any of the many mechanical devices that could break down there. (Of course, it doesn't operate the Liquor Storage Room.) That key could be considered a Master Key.

There are six offices, each with one employee, and each employee has a key that opens only his or her area. Each of these keys would also be a change key. The Office Manager has a key that opens all six areas. His key is also a **MASTER KEY**.
The Assistant Manager of the Motel has a key that opens all of the doors in the Front Building (except the Liquor Storage Room). His key, which operates the locks from both the Maintenance Engineer's Master Key Group, and the Office Manager's Master Key Group, would be a **GRAND MASTER KEY**, because it operates more than a single Master Key Group.

The Manager of the Motel has a key that opens every door in all the buildings, except the Liquor Storage Area. Because his key is higher than the Head of Housekeeping's great Grand Master Key, and operates locks that the Great-Grand Master Key cannot, his key is called the **GREAT-GREAT-GRAND MASTER KEY**, because it operates more than a single Great-Grand Master Key Group.

NOTE: IN OUR EXAMPLE, THE HOTEL MANAGER'S GREAT-GREAT-GRAND MASTER KEY OPERATES ALL THE LOCKS IN THE WHOLE SYSTEM, EXCEPT FOR ANY SINGLE-KEYED AREAS. (The Liquor Storage room, for example.) That makes the system a **GREAT-GREAT-GRAND MASTER KEY SYSTEM**. The highest level of master key in the system defines the system.

Incidentally, the Hotel Manager's own suite is operated by his Great-Great-Grand Master Key, but by no other keys. We would, therefore, say that the lock is **KEYED DIRECTLY TO** his Great-Great-Grand Master Key. This is accomplished by pinning it to his key's bitting directly, with no master pins.

If it had been a lower level master key, such as the Head of Housekeeping's Great-Grand Master Key, that we were keying it directly to, however, it might have been necessary to add master pins to the lock to allow his higher level of master key to still operate it as well. Otherwise it would be considered to be **KEYED DIRECTLY TO THE GREAT-GRAND MASTER KEY (NMK).**

If it had its own key as well, but did not fall under the control of any of the lower levels of master key, we would say that it was **UNDER THE DIRECT CONTROL OF** the Great-Great-Grand Master Key. We could have one lock or multiple areas keyed this way.

Also, it could have been keyed **UNDER THE DIRECT CONTROL OF** a Grand or Great Grand Master Key, instead, in which case there would be no individual master key operating it.)

By the way, we often use the symbol **GM** for Grand Master, **GGM** for Great Grand Master, and **GGGM** for Great-Great-Grand Master Key. Some Master keying Locksmiths use the symbols with the letter K behind them, such as **GMK, GGMK, GGGMK**. Either is equally acceptable. In fact, the abbreviation for Master Key (MK) is almost NEVER written without the K. Change Key can also be abbreviated **CK**.

If it seems like you will be using a lot of G's or Greats to describe the system, take heart. Although it is necessary to recognize these terms, standard practice today is to instead refer to **LEVELS OF KEYING**. This is simplicity itself in a standard **HIERARCHAL** system, where everything fits into logical relationships, change key to master key to grand master key to great-grand master key, etc. The change key is LEVEL ONE. The Master Key is LEVEL TWO. The Grand Master Key is LEVEL THREE. The Great-Grand Master key is LEVEL FOUR. The Great-Great-Grand Master key is LEVEL FIVE. The Great-Great-Great Grand Master key is LEVEL SIX. The Great-Great-Great-Great-Grand Master key is LEVEL SEVEN. It is rare indeed to have a system higher than Level Seven using standard locks and cylinders/cores.

LET'S REVIEW SOME OF THE TERMS WE HAVE SEEN SO FAR.

(We will see more as we go along.) Book definitions will be followed by relevant excerpts from the LIST Council Dictionary, which states:

" **Copyright © , 2005 by the Lock Industry Standards and Training Council.**

All rights reserved.

Permission is hereby granted to reprint terms and definitions contained herein with the following stipulations:

1. Any term and its definition are to be reprinted in their entirety.
2. Credit is to be given to the Lock Industry Standards and Training (LIST) Council."

Basic Terms related to the Customer's System:

BITTING- The combination of cuts on the key, - what cut depths appear in which positions (spaces).

Related LIST Council Dictionary Terms:

bitting depth
n. the depth of a cut which is made into the blade of a key

bitting list
n. a listing of all the key combinations used within a system. The combinations are usually arranged in order of the blind code, direct code, and/or key symbol.

bitting orientation
n. the specification of how a written combination is to be applied to bitting positions of a key (tip-to-bow, etc.)

bitting position
n. the location of a key cut

BLIND CODE- Symbols on the key give no information as to its cut depths, or its position in the system. A list may be available to allow it to be translated.

Related LIST Council Dictionary Terms:

blind code
n. a designation, unrelated to the bitting, assigned to a particular key combination for future reference when additional keys or cylinders may be needed

CHANGE KEY- These are the basic keys of the system, that each control individual locks, or individual areas. They are often referred to as room keys or tenant keys. These control one basic lock combination only, but more than one lock can be set to operate on that combination by using the same size set of tumblers in each lock.

> **Related LIST Council Dictionary Terms:**
>
> **change key**
> n. 1. a key which operates only one cylinder or one group of keyed alike cylinders in a keying system. 2. any device that is used to mechanically or electronically allow resetting of certain key or combination locks. see also "reset key" #1
>
> **CK**
> 1. abb. change key 2. abb. control key

CROSS-KEYING - This is when, IN ORDER TO MEET A CUSTOMER'S NEEDS, we allow certain keys to operate specific locks that they would not normally operate. This USUALLY lowers security in the cylinders involved, and should be avoided if possible, on any but the LEAST critical of areas.

This can be Cross-Keying with or without its own individual operating Key (change key).

(Later we will see that it can also be associated or unassociated, but this is based on the progression of the SPF which we have not demonstrated yet.)

> **Related LIST Council Dictionary Terms:**
>
> **controlled cross keying**
> n. a condition in which two or more different keys of the same level of keying and under the same higher level key(s) operate one cylinder by design, e.g., XAA1 operated by AA2 (but not XAA1 operated by AB1). Note: This condition could severely limit the security of the cylinder and the maximum expansion of the system when (1) more than a few of these different keys operate a cylinder, or (2) more than a few differently cross keyed cylinders per system are required.
>
> **cross keying**
> n. the deliberate process of combinating a cylinder (usually in a master key system) to two or more different keys which would not normally be expected to operate it together. See also "controlled cross keying" and "uncontrolled cross keying."

uncontrolled cross keying
n. a condition in which two or more different keys under different higher level keys operate one cylinder by design; e.g., XAA1 operated by AB, AB1. Note: This condition severely limits the security of the cylinder and the maximum expansion of the system, and often leads to key interchange.

DIRECT DIGIT CODE- When the bitting appears on the bow of the key or other prominent place (such as on the box).

Related LIST Council Dictionary Terms:

direct code
n. a designation assigned to a particular key which includes the actual combination of the key

DISPLAY KEY- A specially designed key that bypasses the Hotel function in the lockset(s) for a particular room. It is issued to guests with special security needs while in the Motel. A spanner key sets the Hotel function in place semi-permanently. The DISPLAY Key becomes the ONLY operating key for the room until the spanner key is used to return the lockset to normal position.

Related LIST Council Dictionary Terms:

display key
n. a special change key in a hotel master key system which will allow access to one designated guest room, even if the lock is in the shut out mode. It may also act as a shut out key for that room.

EMERGENCY KEY- (or EMERGENCY MASTER KEY) A specially designed key that by-passes the Hotel function in all the locks in the system.

Related LIST Council Dictionary Terms:

emergency master key
n. a special master key which usually operates all guestroom locks in a hotel master key system at all times, even in the shut out mode. This key may also act as a shut out key.

EMK
abb. emergency master key

GRAND MASTER KEY- A key that operates more than one Master Key group is referred to as a GRAND MASTER KEY.

In a large Motel, this might, for example, open all the locks on three or four floors, each with its own floor master. (The number of floors or areas isn't important, only that there is more than one MASTER combination involved.)

GRAND MASTER KEY GROUP-The group of locks controlled (or operated by) one particular Grand Master Key.

GRAND MASTER KEY SYSTEM-A master key system in which the highest level of master is the Grand Master Key.

GREAT-GRAND MASTER KEY- This is a key controlling more than one GRAND MASTER KEY GROUP. In our example of a large Motel, this might open up two or more groups of several floors each, where the Grand Master Keys open one group of floors each, and the individual floors have their own Master Key to open up all the rooms on that floor, each of which is in turn opened by its own change key.

GREAT-GRAND MASTER KEY GROUP- The group of locks controlled (or operated by) one particular Great-Grand Master Key.

GREAT-GRAND MASTER KEY SYSTEM- A master key system in which the highest level of master is the Great-Grand Master Key.

GREAT-GREAT-GRAND MASTER KEY- This is a key controlling more than one Great-Grand Master Key Group.

GREAT-GREAT-GRAND MASTER KEY GROUP- The group of locks controlled, (or operated by) one particular Great-Great-Grand Master Key.

GREAT-GREAT-GRAND MASTER KEY SYSTEM- A master key system in which the highest level of master is the Great-Great-Grand Master Key.

Related LIST Council Dictionary Terms:

grand master key
n. the key which operates two or more separate groups of locks, which are each operated by a different master key

grand master key system
n. a master key system which has exactly three levels of keying

grand master keyed
adj. of or pertaining to a lock or cylinder which is or is to be keyed into a grand master key system

great grand master key
n. the key which operates two or more separate groups of locks which are each operated by a different grand master key

great grand master key system
n. a master key system which has exactly four levels of keying

great grand master keyed

adj. of or pertaining to a lock or cylinder which is or is to be keyed into a great grand master key system

great great grand master key
n. the key which operates two or more separate groups of locks which are each operated by different great grand master keys

great great grand master key system
n. a master key system which has five or more levels of keying

great great grand master keyed
adj. of or pertaining to a lock or cylinder which is or is to be keyed into a great great grand master key system

GGGMK
abb. great great grand master key

GGGMK'D
abb. great great grand master keyed

GGM
abb. great grand master key

GGMK
abb. great grand master key

GGMK'D
abb. great grand master keyed

GM
abb. grand master key

GMK
abb. grand master key

GMK section
abb. grand master key section

GMK'D
abb. grand master keyed

HOTEL FUNCTION LOCKSET-A lockset that allows the tenant, from inside the room, to lock out all standard keys, even the various levels of master keys. (These are becoming quite rare.)

Related LIST Council Dictionary Terms:

hotel cylinder
n. a cylinder which can function to prevent the operation of all keys except emergency, display and some shut out keys. It may also contain an indicator.

KEYED ALIKE- The same combination of pin tumblers in one lock as in another results in the exact same key or keys operating both.

Related LIST Council Dictionary Terms:

keyed alike
adj. of or pertaining to two or more looks or cylinders which have or are to have the same combination. They may or may not be part of a keying system

keyed different
adj. of or pertaining to a group of locks or cylinders, each of which is or is to be combinated differently from the others. They may or may not be part of a keying system.

keyed random
adj. of or pertaining to a cylinder or group of cylinders selected from a limited inventory of different key changes. Duplicate bittings may occur.

KA
abb. keyed alike

KA1, KA2, etc.
symbol which indicates that all cylinders so designated are or are to be operated by the same key(s). The numerical designation indicates the keyed alike group or set.

KA/2, KA/3, etc.
symbol used to indicate the quantity of locks or cylinders in keyed alike groups. These groups are usually formed from a larger quantity; e.g., 30 cylinders KA/2.

KEYED DIRECTLY TO- This term is applied to locks that are operated by no keys lower than a particular Master Key, Grand Master Key, Great-Grand Master Key, Great-Great-Grand Master Key, etc. This is usually done by pinning DIRECTLY to the bitting of that master, as though it were the bitting for an individual key.

Related LIST Council Dictionary Terms:

master keyed only
adj. of or pertaining to a lock or cylinder which is or is to be combinated only to a master key

LEVEL ONE - Lowest Level-individual (Change) keys operate one cylinder or one area of KEYED ALIKE cylinders.

LEVEL TWO - The Master Key Level which operates one area where the cylinders are not Keyed Alike.

LEVEL THREE - The Grand Master Key Level in which this key operates more than one Master Key's area of operation

LEVEL FOUR - This Great Grand Master Key operates more than one Grand Master Key's area of operation.

LEVEL FIVE - This Great Great Grand Master Key operates more than one Great Grand Master Key's area of operation.

LEVEL SIX - This Great Great Great Grand Master Key operates more than one Great Great Grand Master Key's area of operation.

LEVEL SEVEN - This Great Great Great Great Grand Master Key operates more than one Great Great Great Grand Master Key's area of operation

Related LIST Council Dictionary Terms:

levels of keying
n. pl. the divisions of a master key system into hierarchies of access, as shown in the following tables. Note: the standard key coding system has been expanded to include key symbols for systems of more than four levels of keying.

(The following chart is excerpted from the LIST Council Dictionary)

TWO LEVEL SYSTEM

LEVEL OF KEYING	KEY NAME	ABB.	KEY SYMBOL
Level II	master key	MK	AA
Level I	change key	CK	1AA, 2AA, etc.

THREE LEVEL SYSTEM

LEVEL OF KEYING	KEY NAME	ABB.	KEY SYMBOL
Level III	grand master	GMK	A
Level II	key		
Level I	master key	MK	AA, AB, etc.
	change key	CK	AA1, AA2, etc

FOUR LEVEL SYSTEM

LEVEL OF KEYING	KEY NAME	ABB.	KEY SYMBOL
Level IV	great grand master	GGMK	GGMK
Level III	key		
Level II	grand master key	GMK	A, B, etc.
Level I	master key	MK	AA, AB, etc.
	change key	CK	AA1, AA2, etc.

FIVE LEVEL SYSTEM

LEVEL OF KEYING	KEY NAME	ABB.	KEY SYMBOL
Level V	great great grand master	GGGMK	GGGMK
Level IV	key		
Level III	great grand master key	GGMK	A, B, etc.
Level II	grand master key	GMK	AA, AB, etc.
Level I	master key	MK	AAA, AAB, etc.
	change key	CK	AAA1, AAA2, etc.

SIX LEVEL SYSTEM

LEVEL OF KEYING	KEY NAME	ABB.	KEY SYMBOL
Level VI	great great grand master key	GGGMK	GGGMK
Level V	great grand master key	GGMK	A, B, etc.
Level IV	grand master key	GMK	AA, AB, etc.
Level III	master key	MK	AAA, AAB, etc.
Level II	sub-master key	SMK	AAAA, AAAB, etc.
Level I	change key	CK	AAAA1, AAAA2, etc.

MAISON KEYING – When all the keys from an area such as an apartment complex pass a specific door(s) or cylinder(s) it is often referred to as Maison Keying, from the French "maison" for house. Older texts may use the less modern "maisson" which designated more of a mansion than a house, but this is considered archaic (or even totally incorrect) today.

> **Related LIST Council Dictionary Terms:**
>
> **Maison key system**
> n. [from the French, meaning "house" key system] a keying system in which one or more cylinders are operated by every key (or relatively large numbers of different keys) in the system; e.g., main entrances of apartment buildings operated by all individual suite keys of the building

MASTER KEY- (noun) This is a key which operates more than one change key combination.
For example, in a typical Motel, the room keys would be change keys, but the key to open all the rooms on a particular floor would be referred to as the floor MASTER.

MASTER KEY- (verb) To cause or develop a system as to cause certain keys to be capable of controlling or operating locks which are not KEYED ALIKE.

(adjective) descriptive of the techniques used to describe such a system, or to describe the system itself.

SPECIAL NOTE: At one time it was common practice to use "Masterkey" for the verb to avoid confusion with the "Master Key" (noun), which is the key created to operate all or part of such a system. This has fallen from usage within the trade and is no longer considered correct.

Related LIST Council Dictionary Terms:
master key
n. 1. a key which operates all the master keyed locks or cylinders in a group, each lock or cylinder usually operated by its own change key, 2. v. to combinate a group of locks or cylinders such that each is operated by its own change key as well as by a master key for the entire group, 3. n. an automotive key which operates all or most locks on a vehicle where there is also a valet key employed
Related LIST Council Dictionary Terms:

MK
abb. master key

MASTER KEY GROUP- The group of locks controlled (or operated by) one particular Master Key.

Related LIST Council Dictionary Terms:

master key changes
n. the number of different usable change keys available under a given master key

MASTER KEYED – pertaining to a lock or group of locks which have been prepared to accept at least two levels of keying, e.g. change key and master key.

Related LIST Council Dictionary Terms:

master keyed
adj. 1. of or pertaining to a cylinder or group of cylinders which are or are to be combinated so that all may be operated by their own change key(s) and by additional key(s) known as master key(s) 2. a warded lock whose warding has been designed to allow different hierarchies of keys to operate

MK'd
abb. master keyed

MASTER KEYING- The creation of a carefully planned system of key based Controlled Access, on a "NEED TO GO" basis.

NO MASTER KEY (NMK) – A specially prepared cylinder combination that used elsewhere within the system as a standard bitting, but is used in this specific cylinder as a Single Keyed combination, with no higher master keys operating it.

> **Related LIST Council Dictionary Terms:**
>
> **NCK**
> sym. symbol for "no change key," primarily used in hardware schedules
>
> **NMK**
> abb. a notation used to indicate "not master keyed" and is suffixed in parentheses to a regular keying symbol. It indicates that the cylinder is not to be operated by the master key(s) specified in the regular keying symbol; e.g., AB6(NMK).

SHUT-OUT KEY- A specially designed key that bypasses the Hotel function in the lockset(s) for a particular room. It is issued to guests with special security needs while in the Motel. A spanner key sets the Hotel function in place semi-permanently. The Shut- Out Key becomes the ONLY operating key for the room until the spanner key is used to return the lockset to normal position.

> **Related LIST Council Dictionary Terms:**
>
> **shut out key**
> n. usually used in hotel keying systems, a key which will make the lock inoperative to all other keys in the system except the emergency master key, display key, and some types of shut out keys
>
> **shut out mode**
> n. the state of a hotel function lockset which prevents operation by all keys except the emergency master key, display key, and some types of shut out keys

SIMPLE MASTER KEY SYSTEM- A system in which there is only one level of master key, and only one bitting for it. The number of change keys under it can be any number greater than two.

SINGLE-KEYED- (also called **SINGLE-KEYED DIFFERENT**) This term applies to locks which are not opened by any Master Key, Grand Master Key, Great-Grand Master Key, Great-Great-Grand Master Key, etc. Even though These keys might, at first glance, appear not to be a part of the system, since no master key of any level operates them, in reality they are an important part of them. This is because we must be absolutely certain that no keys in the system will accidentally open it, nor that its keys would accidentally operate any of the master keyed cylinders.

SPANNER KEY- A key with two small protrusions at the tip of it to fit into similar holes in the push button of a Hotel function Lockset to turn it and engage or disengage the function semi-permanently.

UNDER DIRECT CONTROL OF- Used to describe locksets whose control skips lower level master(s) to be controlled by a higher level master.

VISUAL KEY CONTROL- Symbols on the key telegraph its position in the system to knowledgeable persons.

Related LIST Council Dictionary Terms:

visual key control
n. a specification that all keys and the visible portion of the front of all lock cylinders be stamped with standard keying symbols

The system we created for our fictional example (No, Virginia, there isn't REALLY a "Cost-Moore Motel"...) was more complex than are typical systems in use today. It allowed us to see all the terms in use (in action, so to speak) however, which should help in remembering their meanings.

Now it is time to introduce you to terms related to one of the simpler tools of the Master keying Locksmith....

Lesson Two:
The Key System Schematic and SKCS

The **KEY SYSTEM SCHEMATIC** (formerly known as a **Flow Chart**) is a flat picture of the master key system that, at a glance, shows you the entire system--what locksets are controlled by which key(s), and where Single-Keyed or Cross-Keyed locks exist. Master Keys are identified, and their areas of authority are defined.

> **Related Terms from the LIST Council Dictionary:**
>
> **key system schematic**
> n. a drawing with blocks utilizing keying symbols, usually illustrating the hierarchy of all keys within a master key system. It indicates the structure and total expansion of the system.

To do this, the KEY SYSTEM SCHEMATIC uses rectangular blocks of varying sizes to identify Master Keys of varying levels, and smaller blocks or ovals to represent change keys or the locks they operate. A recent practice is to use ovals for change keys, because the use of ovals more readily calls attention to them, and further separates them from Master Keys. Either the small rectangles OR the ovals is considered technically correct, however.

The blocks or ovals are connected by "**LINES OF CONTROL**" leading from the top of the chart, where the highest master key in the system is shown, down to each lower level of master, and from it to still lower levels, and eventually down to the change key area, where each change key comes off horizontally from a vertical line.

Single-Keyed locks are shown as a separate rectangle or oval, and to identify their individuality, there are usually no lines of control connected to them.

Cross-Keyed locks are shown at the level that best shows their position in the system, usually with arrows attached to the lines of control to further clarify them.

If all that sounds complicated, the following illustrations should help clear it up.

SIMPLE SYSTEM

To start our illustrations of the **FLOW CHART**, we'll look at a chart for a **SIMPLE MASTER KEY SYSTEM**, (**TWO LEVEL**) where only one level of master key exists.

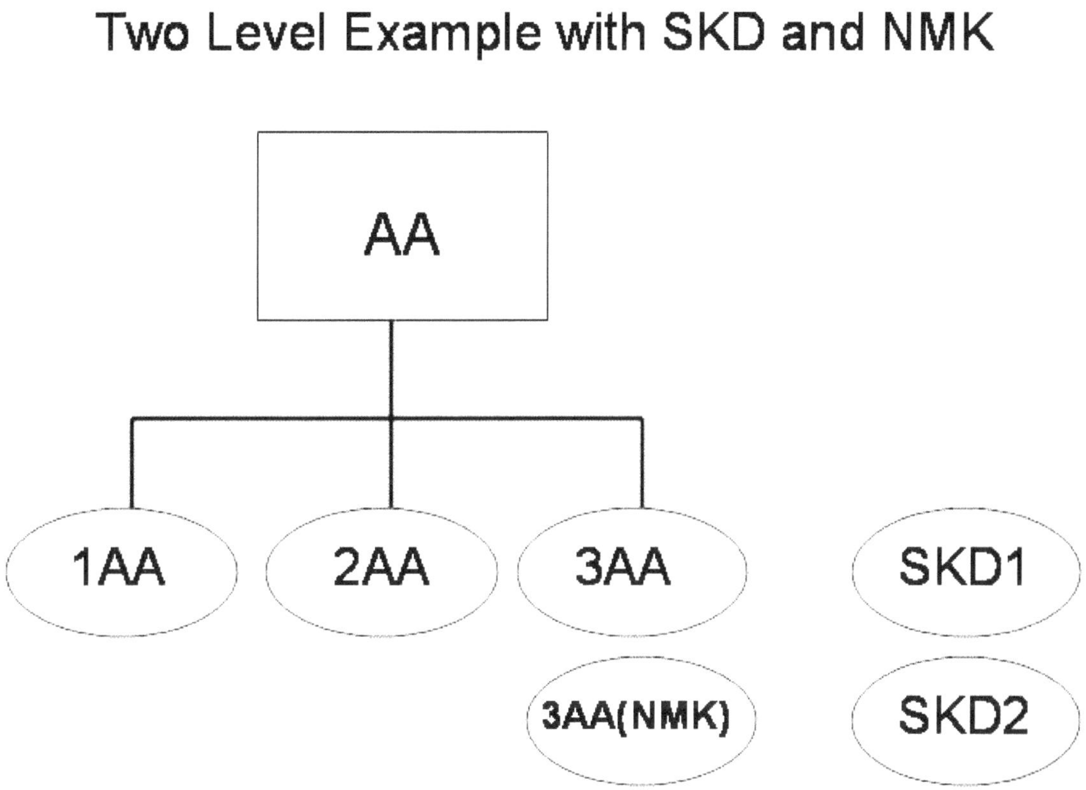

(The symbols on the chart are from the Standard Key Coding System (SKCS), which we will look at later.)

The change keys marked 1AA, 2AA, and 3AA could have been shown vertically instead of horizontally. Either is correct. Use the one that fits the situation the best. **They can also be grouped such as change keys 1AA to 246AA being in a single oval.**

The next illustration shows a **GRAND MASTER KEY SYSTEM** (**3 Level**) where the Grand Master Key is the highest level of key, and master keyed areas exist below it):

Three Level System Example with SKD and NMK

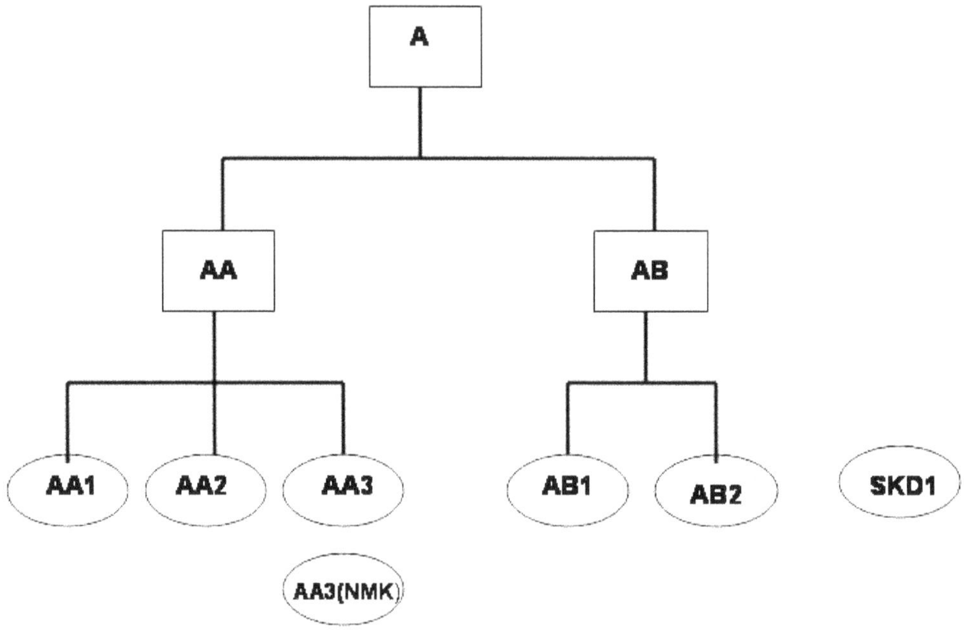

And finally, the next illustration shows a **GREAT-GRAND MASTER KEY SYSTEM**, (where the highest level of master is the Great Grand Master, and more than one Grand Master Key Group exists, each with more than one Master Key Group under it:

Four Level System Example with SKD and NMK

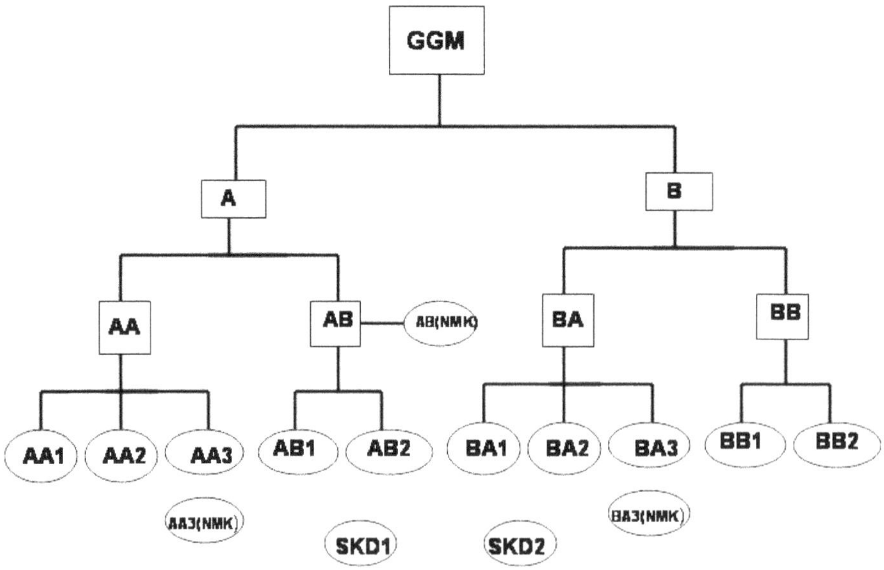

And now for one of the most useful tools at the Master keying Locksmith's Disposal...

STANDARD KEY CODING SYSTEM

STANDARD KEY CODING SYSTEM is the Master keying Locksmith's shorthand. It is used by virtually every lock manufacturer, architect, Architectural Hardware Consultant, and other persons professionally involved in master keying. Unfortunately, too many locksmiths have never really been correctly introduced to this valuable tool.

One thing to understand is that it was intended to depict your customer's system, including anticipated expansion needs, NOT every key that could POSSIBLY be added to the system. This is an important distinction, because not every key on a SPF chart will FIT into it!

Some locksmiths, not understanding this simple fact, have pressured software developers into adding it to the Master Keying software on the market. This results in either not including the standard options for the Master Key charts or coming up with "jokers" ("joker is wild") – labels that have no acceptance in the trade, usually for a good reason. The SKCS (sometimes written as the SKC System or incorrectly written as SKC) loses much of its value when jokers are added to the mix. Without them, it is one of the most useful tools for system development and system maintenance.

Related Terms from the LIST Council Dictionary:

standard key coding system
n. an industry standard and uniform method of designating all keys and/or cylinders in a master key system. The designation automatically indicates the exact function and keying level of each key and/or cylinder in the system, usually without further explanation.

SKCS
abb. standard key coding system

STANDARD KEY CODING SYSTEM has also been called **Uniform Keying Designation** or **Visual ID**, or when applied to the locks and keys by stamping them, as **VISUAL KEY CONTROL**. It is in your best benefit to recognize all the applicable terms. When applied to the system layout but not the face of the cylinders, cores or keys, it can be called **CONCEALED KEY CODING**.

Related Terms from the LIST Council Dictionary:

visual key control
n. a specification that all keys and the visible portion of the front of all lock cylinders be stamped with standard keying symbols

concealed key control
n. a specification that all keys and the non-visible portion of all lock cylinders be stamped with standard keying symbols

The original Standard Key Coding System was developed by Tom Hennessey, a locksmith employed by the Corbin lock company at the time, and was intended to be an in-house tool to give the company an edge. But it was just too efficient and was soon adopted by Yale lock company at which time word of it spread throughout the industry.

The LIST (Lock Industry Standards and Training) Council Dictionary lists an additional definition relating directly to this fine tool:

Related Terms from the LIST Council Dictionary:

modified key coding system
n. a modification of the standard key coding system, in which the TMK of a system with three or more levels of keying has a single letter name, and keys at lower levels are named following SKCS conventions

Note that it was necessary to state that the system had continued to evolve from its original form. This is true even today. What was once the "Modified Key Coding System" has continued to change to better fit the industry as a whole. Today, what is accepted as the SKCS or Standard Key Coding System is recognized as a "living document" that will continue to change as need be.

Simply put, what happens with **STANDARD KEY CODING SYSTEM** is that letters are substituted for master keys (of various levels) and a combination of letters and numbers is used to identify change keys. This is done in a simple and straight-forward way that allows you to see what locks are controlled by which keys with only a quick look. It also fits very nicely with

There are three basic variations involved with **STANDARD KEY CODING SYSTEM**, and these are based on the complexity of the system. The basic system is designed around a GREAT GRAND MASTER KEY SYSTEM, (where the highest level of master key is the GGMK), but it can also be applied WITHOUT CHANGING THE RULES OF IT to a GRAND MASTER KEY SYSTEM.

The second variation is for the **SIMPLE MASTER KEY SYSTEM**, where only one level of master key exists (a two level system – change keys and master key), and there are no Grand Master Keys. The system changes here to call attention to the absence of any Grand Master or higher keys.

The third variation is for **GREAT-GREAT GRAND MASTER KEY SYSTEMS** or higher where there are more than one Great Grand Master Key, and it is useful to be able to identify which is which.

Now let's review the terms and usage of the Standard Key Coding System (SKCS)

Standard Key Coding System

(Also known as Uniform Keying Designation)

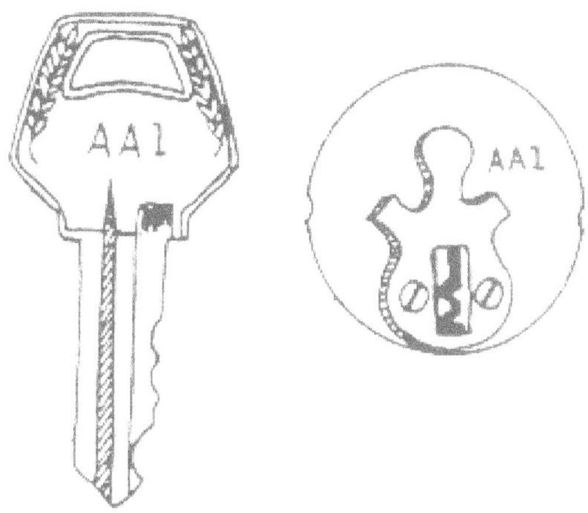

Master Key Systems May Be:

- **Hierarchy Oriented:** There is a Top-Down hierarchy whereby "Higher" Level Keys operate more areas in the system.
- **Function Oriented:** There are fewer Levels in the system and they are based on the usage of the area.

While the standard key coding system provides usable symbols for both types of systems, it is generally more effective and useful a tool in Hierarchy Oriented systems, so we will examine its application for that type first, following a review of some simple terminology.

Standard Hierarchy Terms:

- *Level One*: ***Lowest Level-individual (Change) keys operate one cylinder or one area of KEYED ALIKE cylinders.***

- *Level Two*: ***The Master Key Level which operates one area where the cylinders are not Keyed Alike.***

- *Level Three*: ***The Grand Master Key Level in which this key operates more than one Master Key's area of operation***

- *Level Four*: ***This Great Grand Master Key operates more than one Grand Master Key's area of operation.***

- *Level Five*: ***This Great Great Grand Master Key operates more than one Great Grand Master Key's area of operation.***

- *Level Six*: ***This Great Great Great Grand Master Key operates more than one Great Great Grand Master Key's area of operation.***

- *Level Seven*: ***This Great Great Great Great Grand Master Key operates more than one Great Great Great Grand Master Key's area of operation.***

In a hierarchy based system using standard lock cylinders, a level seven is as high as your system is likely to be capable of becoming.

In specialty products, it may be capable of development beyond that point or it may, in fact, be more limited.

2 Level System

- ***General system has 1 Master that works all or most of the changes.***

- ***Masters are given just letters as designations.***

2 Level System

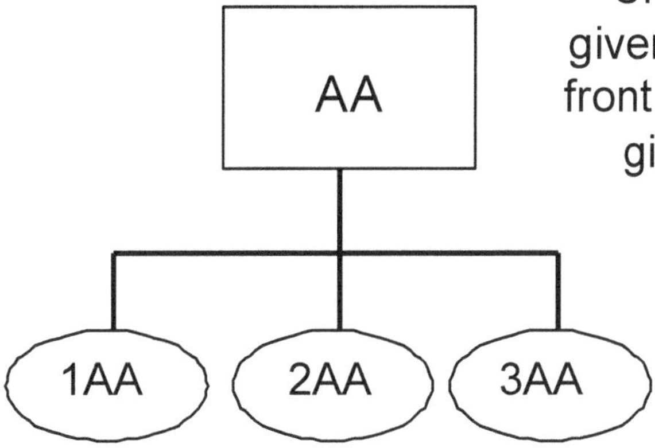

Changes are given numbers in front of the letters given to the master

2 Level System

•Changes not to be operated by the master are given a "Single Keyed" (SKD) label.

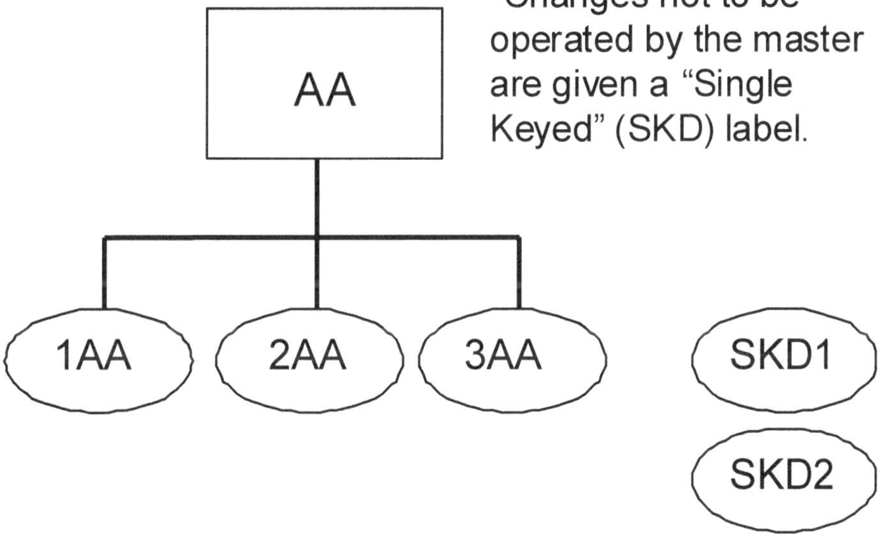

SKD and NMK

Single Keyed (SKD)

- Part of the system
- Same keying may be used in more then one lock
- Never intended to be masterkeyed
    - SKD1

Not Master Keyed (NMK)

- Part of the system
- Same keying may be used in more then one lock
- Other locks with same bitting intended to be masterkeyed - except the lock(s) given the (NMK) label
    - AA1(NMK)

Two Level Example with SKD and NMK

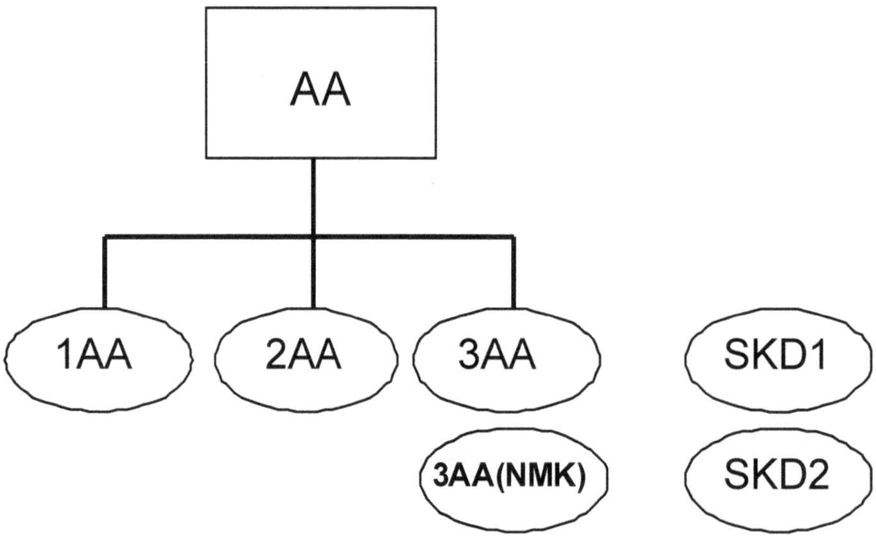

3 Level System

- General system has 1 Master that works all or most of the changes
- Masters are given just letters as designations.

3 Level System

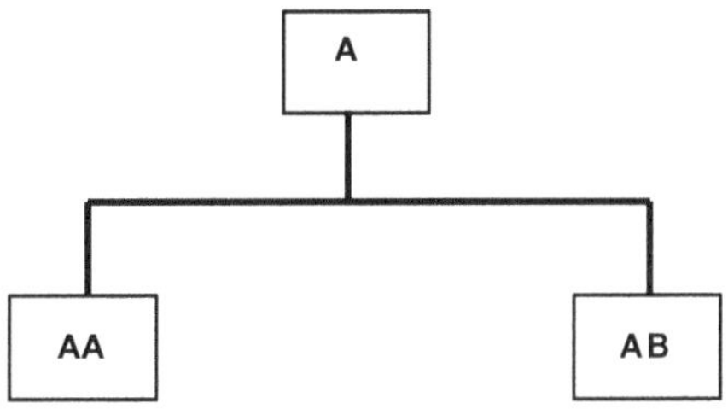

•Masters are given a letter behind the letters given to the Grand Master

3 Level System

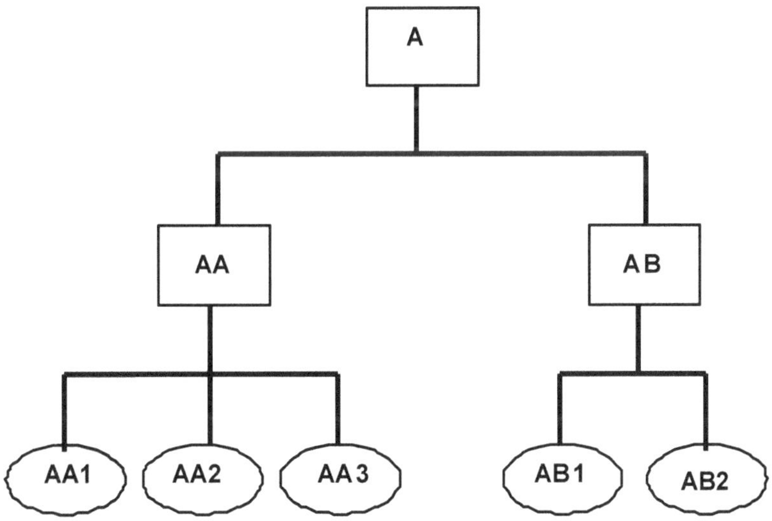

•Changes are given numbers behind the letters given to the master.

Three Level System Example with SKD and NMK

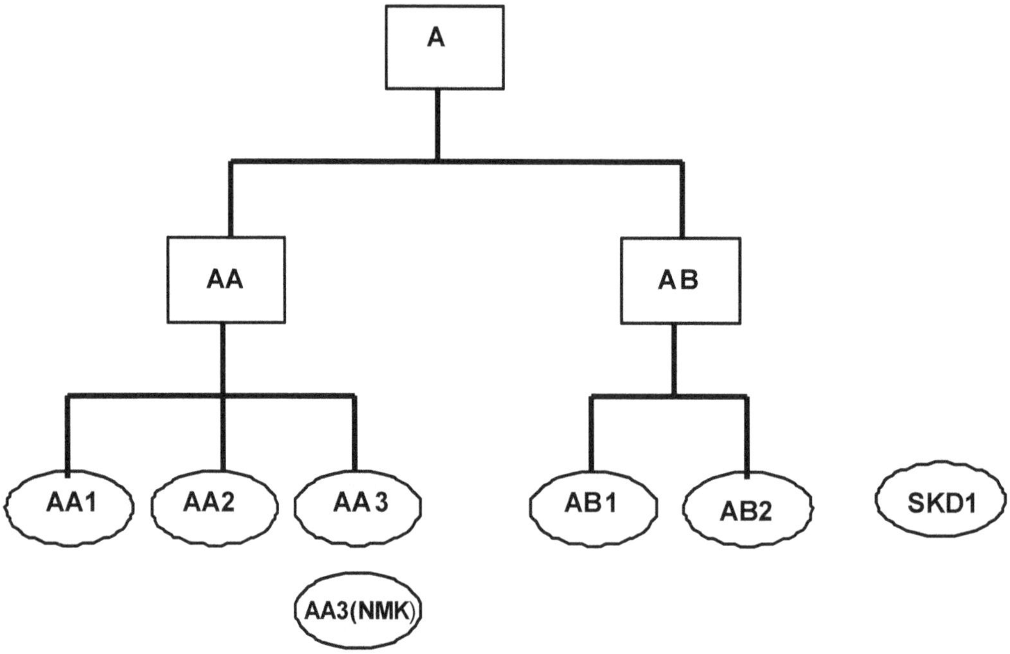

4 Level System

- General system has 1 Great Grand Master that works all or most of the lower masters and changes. Great Grand Master is given the label (GGM)

- Grand Masters are given a single letter as designation.

- Masters are given a letter behind the letters given to the Grand Master above it.

- Changes are given numbers behind the letters given to the master it falls under.

- SKD : These are handled the same as under level 2 system.

NMK : AA1(NMK) operated by change AA1 only
AA(NMK) operated by Master AA only
A(NMK) operated by Grand Master A only

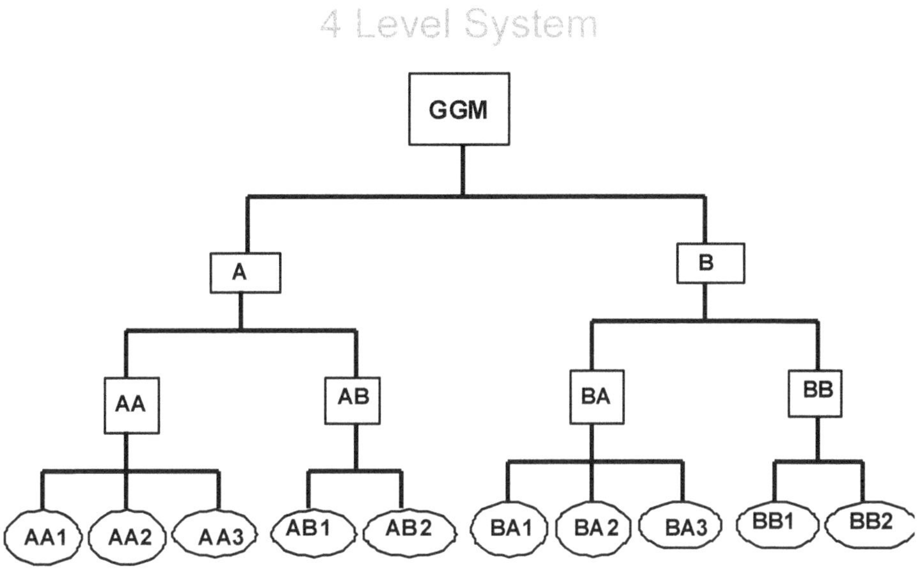

•Changes are given numbers behind the letters given to the master it falls under.

When you have a change key that you want to be operated by just a Great Grand Master or Grand Master but not a lower Master you give it a label of the Great Grand Master or Grand Masters letter and number(s)

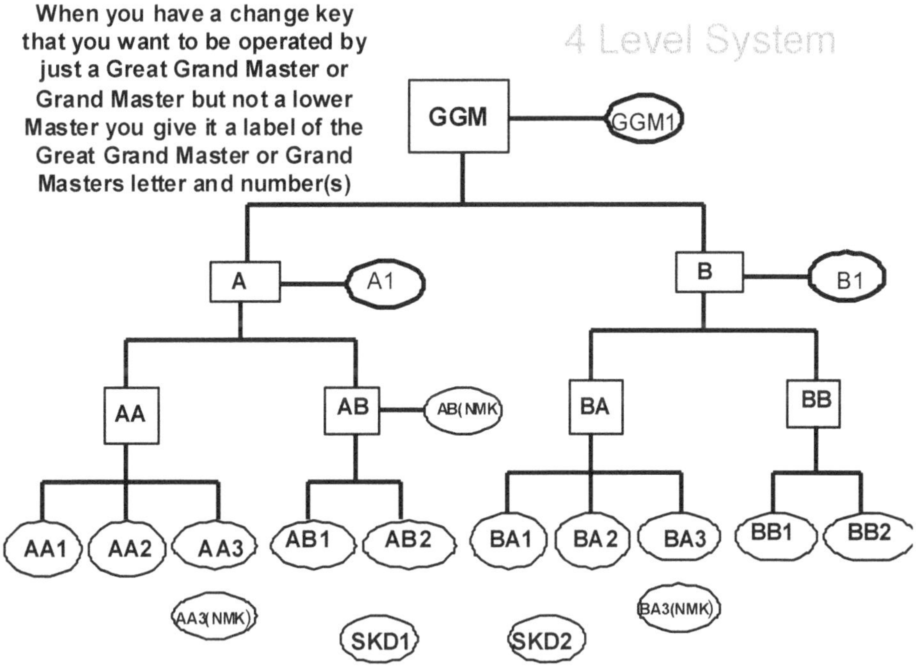

5 Level System

- ***General system has 1 Great Great Grand Master that works all or most of the lower masters and changes. Great Great Grand Master is given the label (GGGM)***
- ***Great Grand Masters are given a single letter as designation.***
- ***Grand Masters are given a letter behind the letters given to the Great Grand Master above it.***
- ***Masters are given a letter behind the letters given to the Grand Master above it.***
- ***Changes are given numbers behind the letters given to the master it falls under.***

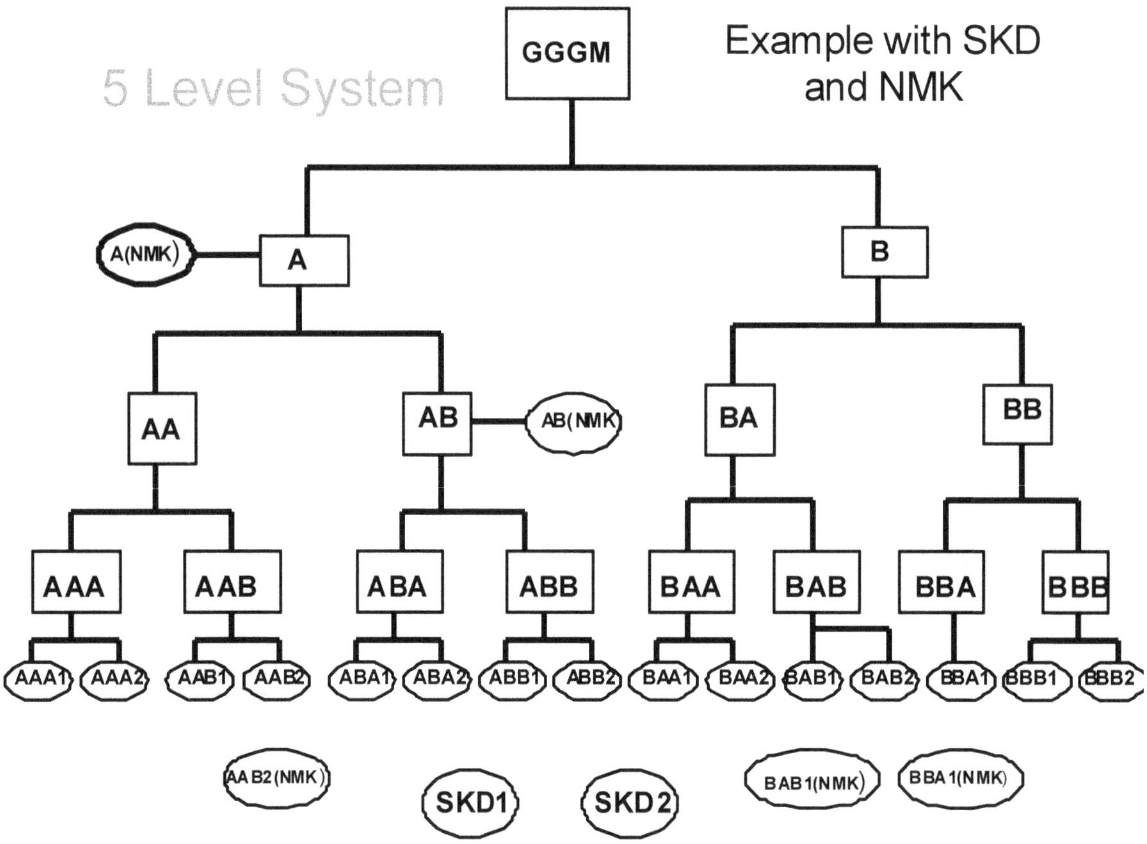

Special Situations

No Grand Master Key (2 level): The master key typically is called **AA**, and the change keys are **1AA, 2AA, 3AA**, etc. The numbers are **PREFIXED TO** (added to the front of) the master key symbol. This is a Level 2 System.

KEYED DIRECTLY TO: A cylinder or cylinders are operated by no keys lower than the specified master key level. Standard coding takes the symbol of that Master Key Level, such as A for Keyed Directly To Grand Master Key.

KEYED DIRECTLY UNDER: A cylinder or cylinders are operated by individual keys and the specified Master Key level, but by no intermediate levels. These take the symbol of the specified Master Key Level plus a number **Suffixed to** it, (A1,A2, B1)

SINGLE KEYED: No Master Key of any level operates this (group of) cylinder(s). Takes the symbol SKD suffixed by a number, such as SKD1,SKD2

NMK: No Master Key (Another less common designation for SKD)

CROSS KEYED WITH ITS OWN KEY: Prefixed with X in Front of Standard Symbol for its key (XAA1, XBC5, etc.)

CROSS-KEYING WITHOUT ITS OWN KEY: Pair of X symbols separated by Number (X1X, X2X,etc.)

The following chart shows a five level system with several of the special situations charted. You should be able to identify each.

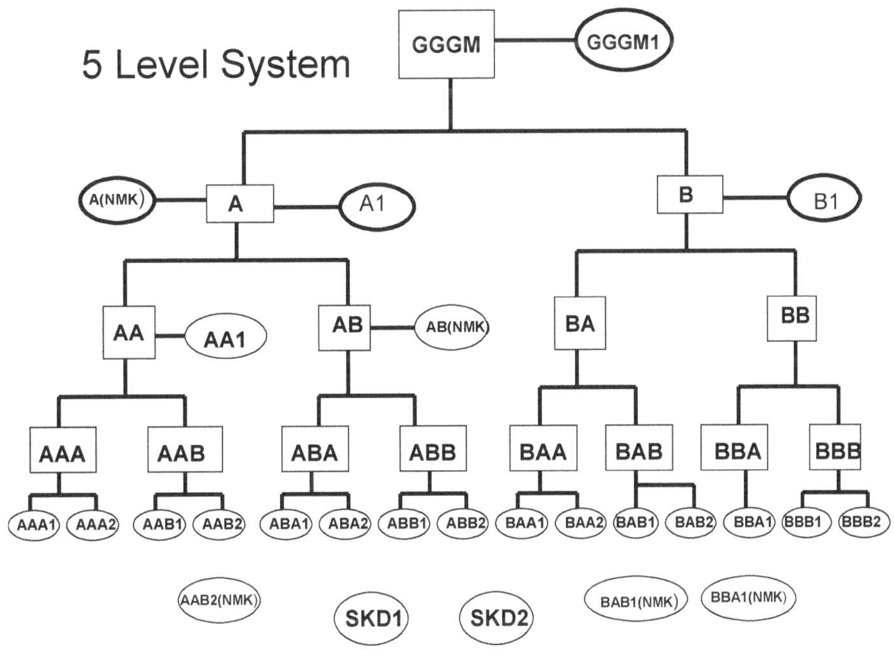

Cross Keying

In some circumstances you may have two or more change keys that have to pass a given lock. This is called Cross Keying. A cross keyed lock can have it own key or not.

Cross keying may be considered Controlled Cross-Keying or Uncontrolled Cross Keying. These terms may be a bit deceptive, because they have nothing to do with whether you intended it or not. "Cross-Keying" that you did not intend to happen is called INTERCHANGE and is a bad thing.

CONTROLLED CROSS KEYING: a condition in which two or more different keys of the same level of keying and under the same higher level key(s) operate the cylinder by design; e.g., XAA1 operated by AA2 (but not XAA1 operated by AB1)

- ***UNCONTROLLED CROSS KEYING:***a condition in which two or more different keys under different higher level keys operate one cylinder by design; e.g., XAA1 operated by AB, AB1

Cross Keying May also be with or without its own key:

Cross Key with its own key

- AA1 works the main door but not the inside offices.
- AA2, AA3 and AA4 work their own offices and the main door

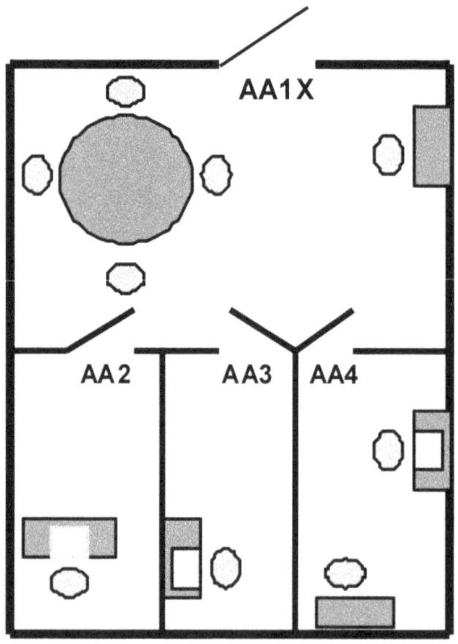

Cross Key with out its own key

- AA1, AA2 and AA3 all work their own rooms but also all work X1X also.

- There is no individual key for doors X1X.

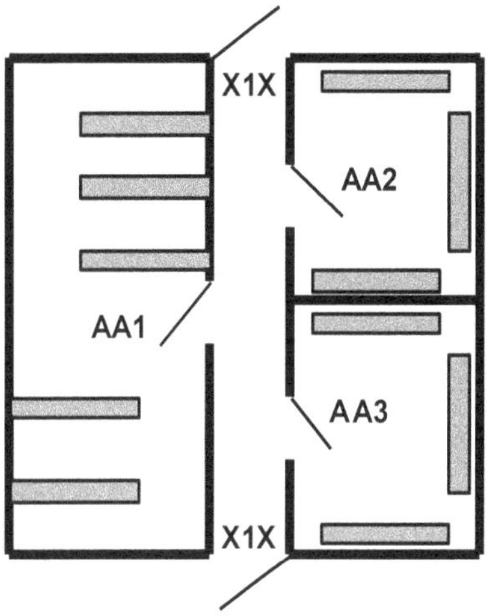

Says ***STOP!*** There is something unique in my keying!

Only by looking at the notes can you see what the unique keying is for an "X" marked lock.

For example, the notes may say:

A1X *operated by AA2, AA3, AA4*

X1X *operated by AA1, AA2, AA3*

AB1X *operated by AB1 through AB64*

An abbreviation used by some manufacturers is:

"O.b" to stand for "Operated by"

For example:

AA1X **O.b AA2, AA3, AA4**

X1X **O.b AA1, AA2, AA3**
AB1X O.b AB1 through AB64

Some Manufacturers also avoid S (5), G, M, and Z (2)

As systems grow you may end up with a large number of masters and changes.

Change keys are simple because you keep increasing the number:

AA1…… AA25… …..….AA120………… …………..AA540

Masters go from:

AA, AB, AC…..AG……………………AZ

After AZ you place a number between the letters:

A2A, A2B, A2C……A2G…………… …….A2Z

On the next pass increase the number to "3"

A3A, A3B, A3C…….A3G…… ……………..A3Z

Function Oriented Systems

Unlike the hierarchy based systems illustrated in the previous pages, in the function oriented system the basic system tends to be very simple and straightforward, with few levels of keying needed.

However, in this type of system, there is a need for certain groups of individuals to access specific types of areas, often not exclusively.

In a typical institution this involves security guards or police, maintenance workers, custodial employees, and certain classes of vendor or contractor.

This typically fits the Selective Master Keying method better than any of the other methods.

SELECTIVE MASTER

- ***A Selective Master is a master that you can place in locks by a case by case basis.***
- ***They are often high level masters related to the Top Master (GGGM, GGM etc.,) but can be left out of locks as needed.***
- ***They are also often related to a need like for Engineering or Housekeeping.***

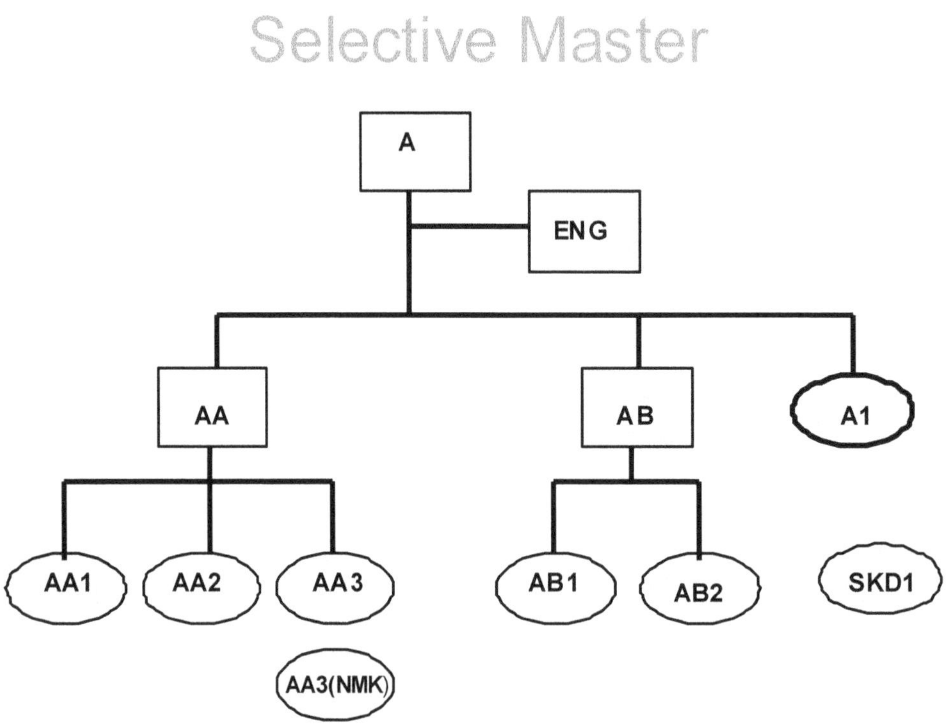

Related Terms from the LIST Council Dictionary: **selective master key** n. an unassociated master key which can be made to operate any specific lock(s) in the entire system in addition to the regular master key(s) and/or change key(s) for the cylinder without creating key interchange

Selective Master Example 1:

5 rooms:
- 3 Keyed Alike
- 2 Keyed different

Inside the end 2 rooms are closets the repair staff has to get into.

The Selective Master (ENG) is keyed in to these 2 rooms but not the other rooms.

The (ENG) key also is the key for the 2 closets.

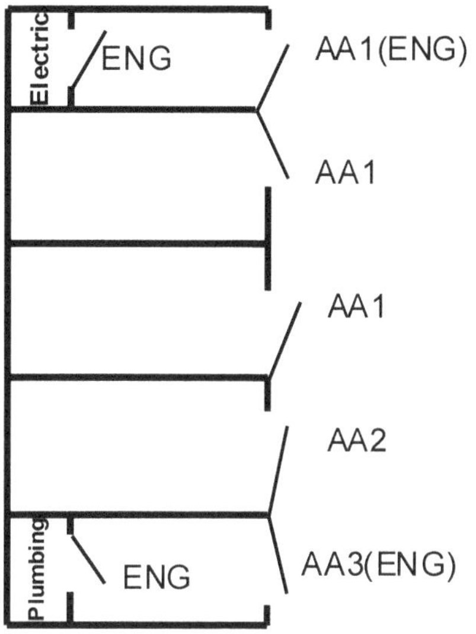

Selective Master Example 2:

5 rooms all keyed different.

Housekeeping has to get in at night to clean but they are not supposed to enter 1 office and the Lab

The (HSK) Selective Master is keyed in to 3 of the locks but not the 2 they are not allowed into.

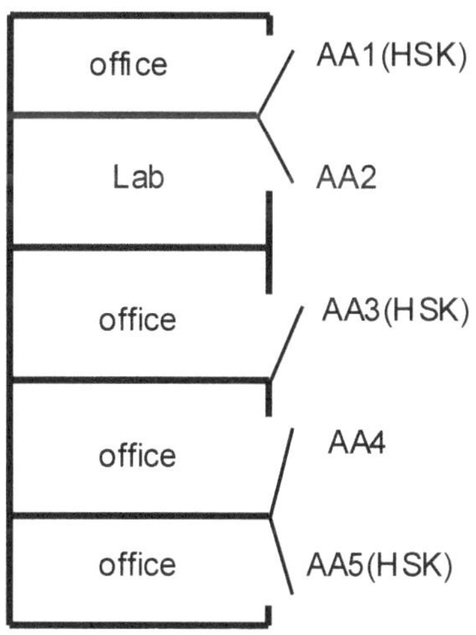

As was mentioned earlier, the SKCS is one of the most useful tools in laying out and identifying needs within the master key system.

However, if we stamp the face of the cylinder with a symbol such as **AA1** and we stamp the matching key with the same symbol, (**VKC-Visual Key Coding**) anyone finding the key can identify where it goes fairly quickly. This is a weakness that we might prefer not to have in our master key systems, so **CONCEALED KEY CODING** (CKC) was developed.

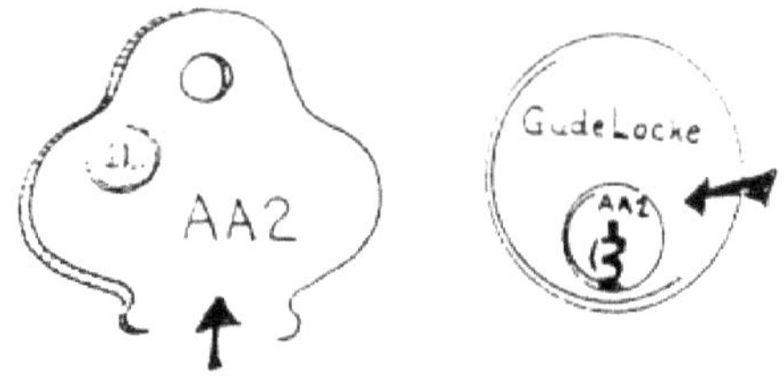

Concealed Key Coding can be as simple as not stamping the face of the cylinder with the SKCS symbols, by placing them in a non-visible location such as the back or side of the cylinder. But if we do that, do we still stamp the key with the SKCS symbol? We can, (The Door and Hardware Institute and the LIST Council both define it in a way that seems to indicate we almost should) but it is not as secure as some other options because if we know that AA2 is the Accounting Room and AA3 is the Accounting paper storage room, we can surmise that AA1 is probably near them as well.

For that reason, it may be desirable to use some other form of key marking. Manufacturers often have their own "joker" specifications (specifications that are recognized by only those associated with their creator rather than the industry as a whole.) to deal with such a situation.

For example, Corbin-Russwin uses the chart on the following page (excerpted from their keying manual).

As you can see, they have covered many options, but not all of the possibilities in these specifications. Many times manufacturers will also have specifications that are reserved for use only within the factory or for special usages. Looking at this chart, it is likely that CKC0 and CKC1 probably fall into this category. But remember, specifications such as these are considered "jokers" (symbols with no industry-wide recognized meaning) and should ONLY be used when dealing with the manufacturer who created and implemented them.

"The following codes and definitions have been established for specifying cylinder and key stamping (by one manufacturer) :

Quick Code Definition

VKC0: No keyset, registry number, keyway, nor bitting stamping on any keys. Includes master keys and change keys. (Keys will be tagged.)
VKC1: Master keys stamped with their keyset symbol and registry number. Change keys stamped with their keyset symbol and keyway.* (Standard for high security and security cylinders.)
VKC2: Master Keys stamped with their keyset symbol and registry number. Change keys stamped with their keyset symbol and keyway.* Cylinders stamped with their keyset symbol on the face of the plug. (Not available for high security or security cylinders.)
VKC3: Master Keys stamped with their registry number. Change keys stamped with their bitting, keyway* and system. Cylinders stamped with their keyset symbol on the face of the plug. (Not available for high security or security cylinders.)
CKC2: Master keys stamped with their keyset symbol and registry number. Change keys stamped with their keyset symbol and keyway.* Cylinders stamped with their keyset symbol in a concealed location.
CKC3: Master keys stamped with their registry number. Change keys stamped with their bitting, keyway* and system. Cylinders stamped with their keyset symbol in a concealed location. (Not available for high security, or security cylinders.)
*** Keyway will NOT be stamped on the keys for security or high security cylinders, nor restricted or patented keyways."**

So what should WE do to not use VKC on the systems we develop? Stamping the cylinder or core in a non-visible portion is easy, but the key is always potentially visible.

We saw earlier that **Direct Code** (formerly known as Direct Digit Code) was when we stamp the keys with the actual bitting (pattern of key cuts), possibly prefixed or suffixed with a symbol identifying the keyway.

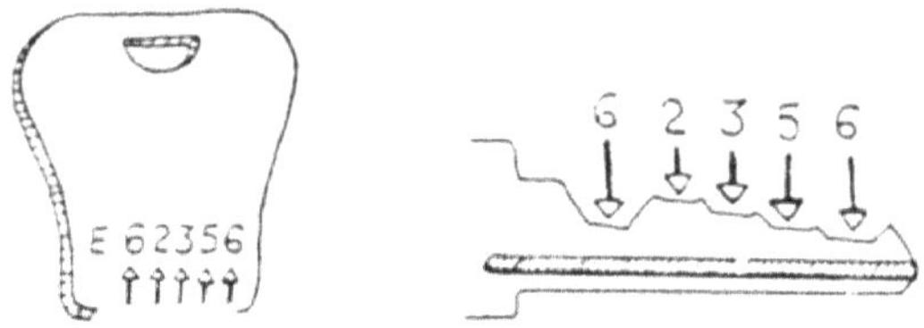

Obviously that is worse than stamping the key with the **SKCS** symbols, because this allows a person with a minimum of knowledge and tools to easily turn one key into another by filing down their key where they see a deeper number on they key they want.

The best option is generally considered to be the usage of a **BLIND CODE**, a code that conveys no information to the average person as to the cut depths or its position within the system.

This can be as simple as making a randomized list of the bittings (pattern of cut depths) that are being used in your system, and assigning random combinations of letters and numbers to them. If you have ever used a code book, you have worked with such a system of symbols.

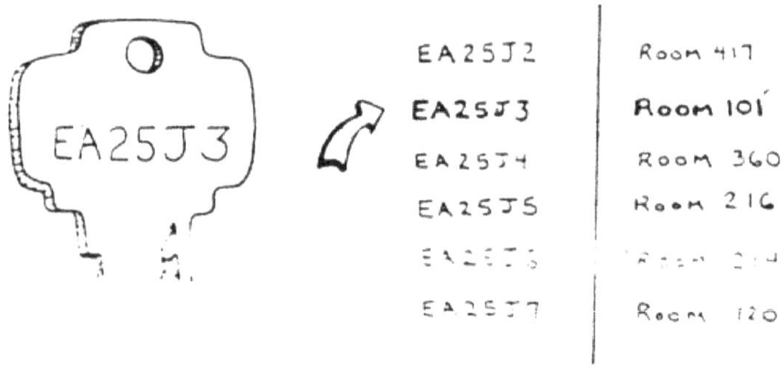

The drawbacks to this simple method are that you need to be careful not to lose this chart or let it be read by anyone else, and that you must be careful not to accidentally duplicate bittings. Computers are great for this, but not so great when you have a system crash, so keep it backed up carefully if you use a computer for it. It is a good idea to keep an "off-site" copy of the backup, but it should be secure and not labeled as such. Labeling it "Photos from Grandma's 80th birthday party" or something of the sort, and the date will probably keep it secure in most cases against prying eyes.

Another simple way of creating a blind code is to serialize the keys. The first key created in 1 followed by 2, etc. The one thousand ninety ninth key created is 1099. Also, you do not have to start at one. The first key could be 998, for example, making the second key 999 and the one thousand ninety ninth 2096.

The drawback to this method is that keys for a given area tend to be created at the same time, causing them to have similar numbers.

Another option is the use of some form of encryption, where the actual bitting is translated into something not so apparent.

There are many other ways to achieve a blind code, and we will cover but a few common ones here. Literally volumes have been written on this topic under the heading of cryptology. Many are available at your local library if this topic begins to fascinate you.

The most common and applicable Blind Codes as shown here are the:

- Direct Transposition Code
- Tabled Transposition Code
- Algorithmic Encryption
- “Key” Encryption

The LIST Council lumps all of these methods of encryption or encoding into one term:

Related Terms from the LIST Council Dictionary:

substitution code
n. a code whose individual characters are converted to individual key cuts or combination numbers by means of a reference table

By simplest definitions, **Direct Transposition Code** is when one letter or number or symbol is substituted for another. Crossword books often feature a section on "cracking" codes like this, which is reasonably easy when dealing with paragraphs because words have patterns that can be used as a weapon against the code. For example, a letter all by itself in a paragraph is most likely an A or I. The larger the sampling, the easier the cracking becomes. However, when dealing with numbers only, no such patterns exist, making the cracking much more difficult.

Tabled Transposition Code is similar, except that instead of being a direct substitution,
a chart is used and what is substituted depends on the position within the code itself.
As such, it is at least ten times as difficult, and possibly more, to crack.

Algorithmic Encryption is using a mathematical formula to decide what is substituted.

"Key" Encryption (not normally surrounded by quotes, but done so here to differentiate it from our purpose of encrypting keys for cylinders and cores) is using an external device to encode/decode.

In the U.S. Civil War (War between the States) soldiers on both sides used this method, with keys varying from the Bible to famous letters, the Declaration of Independence, membership manuals or handbooks in clubs or associations and so forth.

Without the key being known, they are extremely difficult to decode, if not impossible. At Lehigh University in Bethlehem, Pennsylvania, one such encoded letter remains in their Rare Documents area that no one to date has been able to decode or determine the key for.

The LIST Council only currently lists the one term because in most cases it makes no difference to the locksmith how a code was developed. However, if you are going to be developing your own secure blind coding for the keys in your master key system(s), you need to have at least a basic understanding of what your options are and how encoding works.

While it is true that you need not become a full-fledged cryptologist (or generally to defend your system against one) the minimal information contained herein should be considered the least you should know for this purpose!

Of course, the public library in most cities is full of sophisticated texts and basic primers so you can also take this farther than the encoding/decoding techniques shown here, if you decide you need or want to.

Some people use **DERIVATIVE CODES** - codes where the series matches another existing code series (such as for cabinet lock keys or automotive or whatever), but with an additive on one side or the other. This is fairly secure, because SS123 = GEK series code GEK423 (SS+300=GEK) is not very obvious. Still, it is often difficult to find a series that is both complete AND has the exact bittings that you need.

Related Terms from the LIST Council Dictionary:

derived series
n. a series of blind codes and bittings which are directly related to those of another bitting list

DIRECT TRANSPOSITION CODE

A common example of this is the PATHFINDER(S) code series. Although each is well-known, none of its variations telegraphs that this is the method used, so each offers reasonable security.

PATHFINDER(S) uses either the ten digit word **PATHFINDER** or the 11 digit word **PATHFINDERS** as the code key (or translating guide). Basically, a number is substituted for each of the letters, such as:

P=1 A=2 T=3 H=4 F=5 I=6 N=7 D=8 E=9 R=0

Our key from the Direct Code example, with cuts of 62356 would thus be stamped IATFI

The use of the 11th digit, is a code for repeat last number, so that the number 11 for example instead of becoming PP in the first method above, becomes PS, while 22 becomes AS instead of AA, making it even less obvious to the casual observer.

Of course, this could be reversed and be equally simple to remember,

P=0 A=9 T=8 H=7 F=6 I=5 N=4 D=3 E=2 R=1
in which case the example key would be stamped IFTAI

or it could be randomized, (but don't forget your pattern) such as:

P=1 A=0 T=2 H=9 F=3 I=8 N=4 D=7 E=5 R=6
in which case the example key would be stamped RTFER

You could start the transposition at the A and use the P as the repeat last digit

A=0 T=1 H=2 F=3 I=4 N=5 D=6 E=7 R=8 S=9

Want even MORE security? Before encoding it, add a CONSTANT number to the bitting. 451234 + 2345 = 453579 = (standard PATHFINDER) HFTFNE

Because the constant is known only to you, there is no easy way for an outsider to determine that a) the **CONSTANT** was used and b) what the **CONSTANT** is. Just do NOT vary it or you run the risk of forgetting it and being locked out of your own code.

If possible, try NOT to use an easily guessed CONSTANT number such as your phone number or address.

Tabled Transposition

The Basic Chart for "**I James Bond**" is:

	1	2	3	4	5	6	7	8	9	0
1	I	J	A	M	E	S	B	O	N	D
2	J	A	M	E	S	B	O	N	D	I
3	A	M	E	S	B	O	N	D	I	J
4	M	E	S	B	O	N	D	I	J	A
5	E	S	B	O	N	D	I	J	A	M
6	S	B	O	N	D	I	J	A	M	E
7	B	O	N	D	I	J	A	M	E	S
8	O	N	D	I	J	S	M	E	S	B
9	N	D	I	J	A	M	E	S	B	O
0	D	I	J	A	M	E	S	B	O	N

The Basic Chart For **PATHFINDER** is:

	1	2	3	4	5	6	7	8	9	0
1	P	A	T	H	F	I	N	D	E	R
2	A	T	H	F	I	N	D	E	R	P
3	T	H	F	I	N	D	E	R	P	A
4	H	F	I	N	D	E	R	P	A	T
5	F	I	N	D	E	R	P	A	T	H
6	I	N	D	E	R	P	A	T	H	F
7	N	D	E	R	P	A	T	H	F	I
8	D	E	R	P	A	T	H	F	I	N
9	E	R	P	A	T	H	F	I	N	D
0	R	P	A	T	H	F	I	N	D	E

Of course, any ten or eleven digit word or phrase would work the same way for this, and once again, you could add a CONSTANT number before encoding and deduct it after decoding for better security.

Related Terms from the LIST Council Dictionary:

substitution code
n. a code whose individual characters are converted to individual key cuts or combination numbers by means of a reference table

Algorithmic Encryption

In this technique, an algorithm of some sort is applied to the bitting. An algorithm is simply a mathematical expression, such as a+b=c or Log (a)=Int(b) / c^2 or Dec(a)+2140=Hex(b).

Computers use sophisticated 64 bit or 128 bit encryption which is highly secure, but difficult to apply in the field.

A simple algorithm can be highly effective, and yet very easy to code and decode.

The Windows calculator on most computers allows you to **View** it in **scientific** mode. In scientific mode, click the radio button for dec(imal) and enter the bitting. Then click the radio button for hex(adecimal).

The difference is that one is base 10 and one is base sixteen. This creates a shift in both position and value. 234512 becomes 39410, for example, and 329416 becomes 506C8.

To decode it, just reverse the steps. Simple, yet elegant.

You could also use the Oct(adecimal) radio button instead of the Hex(adecimal) on the calculator for another variation. In this case, the 234512 becomes 712020, and the 329416 becomes 1203310.

Ready to convert the cuts? Click the clear (C) button and enter the cut bitting for the key you want to log. (Your radio button is already in DEC –decimal- mode, right?) as shown here for cuts 425103:

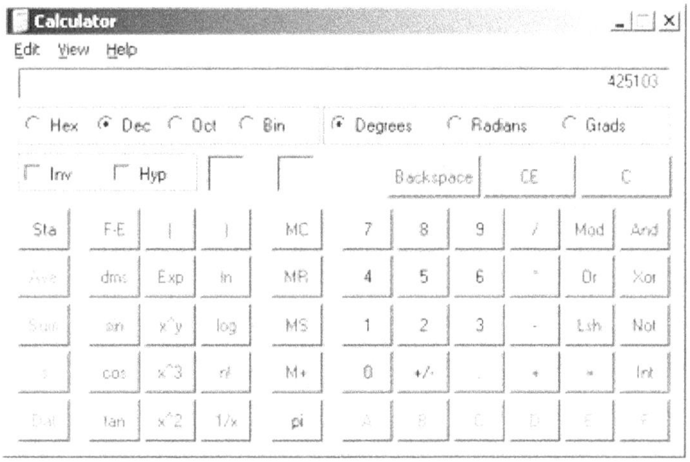

Then click the radio button for HEX (Hexadecimal) or OCT (Octal) to convert it. (We are using Hexadecimal for our example):

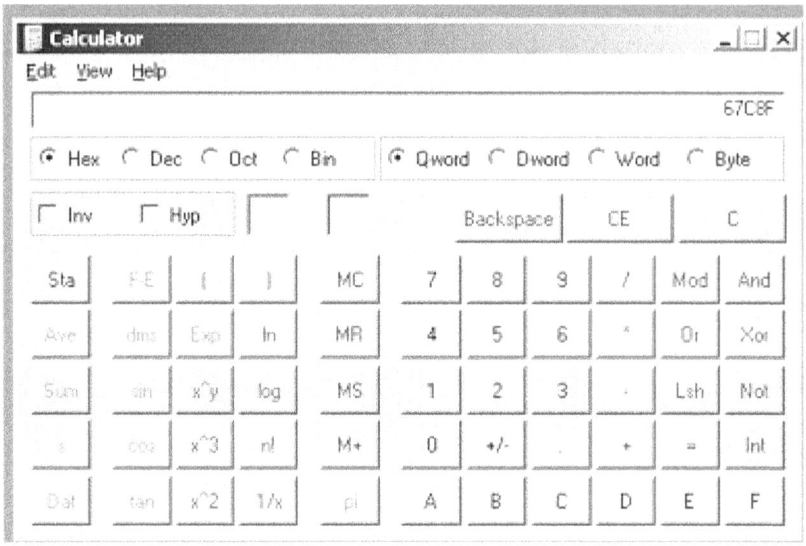

Want even better security? Add a constant number to the bitting before translating it from decimal to hexadecimal or octadecimal. Only YOU know the constant, so it makes no difference if others know this coding technique or not. (This can be done with ANY of the Blind Coding techniques described here.)

Your algorithm here would then be: Dec(a)+b=Hex(c) where a is the starting bitting number, and b is the constant added to it and c is the number to be identified with the key.

You don't need to understand the algorithm to implement it, of course, which is one of the beautiful things about simple, yet elegant, solutions such as this.

“Key” Encryption

The most common form of blind code encryption in use both historically and today is the "key” encryption, in which decoding depends on a knowledge of the “tool” or “key” used to encrypt it in the first place.

Often it is a position in a book or quotation.

Probably the most commonly known of these is the "telephone keypad" code used commonly by teenagers since the 1950's. **The dial portion of the telephone has three letters assigned to each number. For a six digit numeric value there are 729 ways to express it using only the alphabet, and 4.096 ways to express it alphanumerically, yet when decoding it, there is only one possible value for each resulting code.**

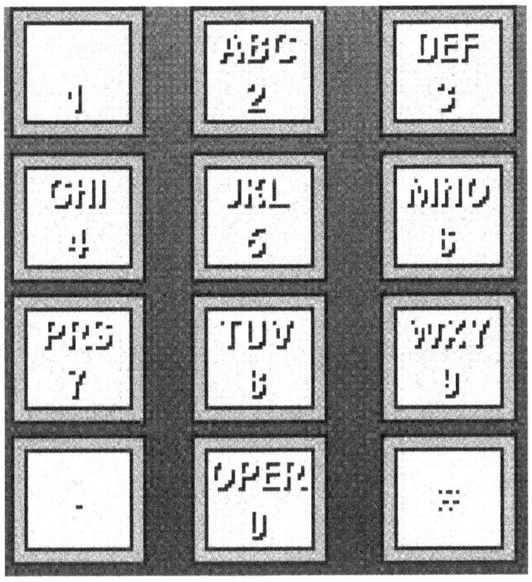

BIDORI = 243674

AGEMSH = 243674

2H3NP4 = 243674

And once again, you could add a CONSTANT to the bitting before encoding it and deduct the constant after decoding it, for better security.

243674 + 2345 = 246019 = (ABC)(GHI)(MNO)01(WXY)

which does not include any of the codes above!

Lesson Three: BASICS OF LOCKS AND KEYS

If you are an experienced lock service person, you can probably skip this lesson.

In a typical pin tumbler cylinder, a key works if two conditions are met. First, it must fit into the keyway of the cylinder, and then once it is in the cylinder, it must raise all the bottom pin tumblers to within the tolerances of a shearline within the lock.

The following illustration shows a series of keyways from one manufacturer. Note the different shapes of the keyways.

The basic shapes that make up a keyway can be:

round, inverse round, V, square,

up (sometimes called Left), and down (sometimes called right).

These can be difficult to see looking into the keyway, but tend to be quite clearly defined on most keys:

If the pattern of grooves on the key matches the pattern of grooves and lands (metal between the grooves) in the keyway, the key will enter the keyway properly.

The proper key raises the bottom pins to the operating shearline. This "opens" the chambers, enabling the plug to rotate, so that the key will turn.

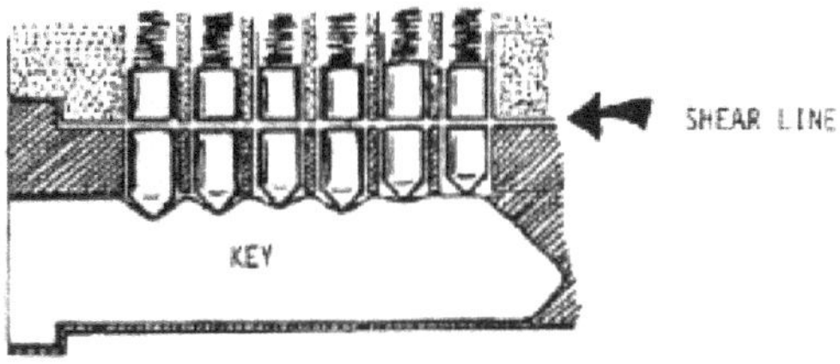

If any of the pins, including the top pins, in any of the chambers are blocking the operating shearline, the plug will not rotate, so the key will not turn.

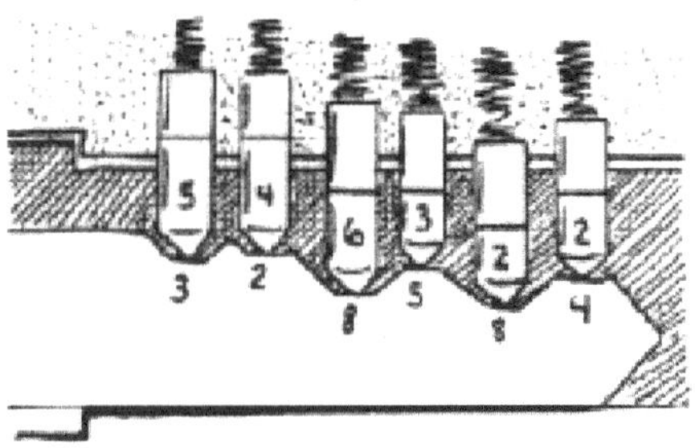

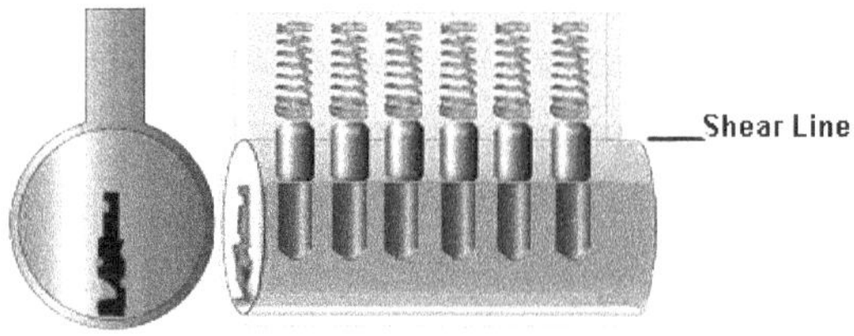

No Key in Cylinder- Plug cannot turn because top pins block the shear line.

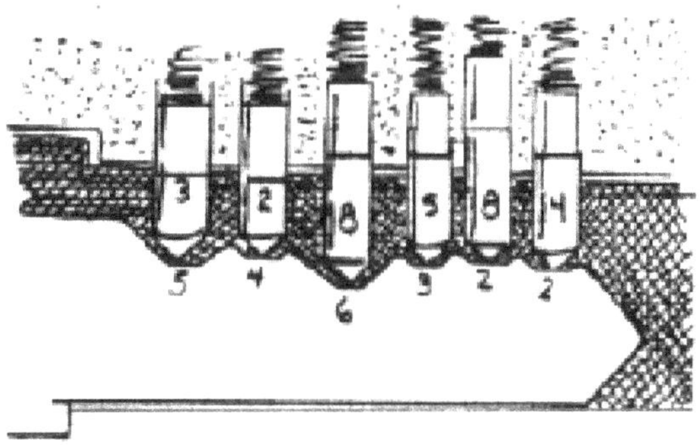

The **328584** and **546322** are examples of the key **BITTING** as we discussed earlier. The cuts are usually measured from the key stop, which may be at the top shoulder, bottom shoulder, the tip of the key or may be a specially designed tip stop cut into the base of the blade of the key at the tip. Most manufacturers read the bitting from the position closest to the stop. There have been and continue to be exceptions to this, however.

Early Corbin keys in the X class key numbering system before 1970 read the bitting from tip to bow (or shoulder) even though they used a shoulder stop. Schlage Everest uses a tip stop but unlike the Best and most other Small Format Interchangeable Core products, reads the cuts from Bow to tip.

Still, more than 98% of the products read the bitting from the cut closet to the stop.

Lock manufacturers choose an increment size that is appropriate for their keys, the size of increments and the number of possible depths. Some use .015" (fifteen thousandths of an inch) for each increment size. Some others use .020" (twenty thousandths of an inch). Others use various other increment sizes. Remember that the tip of the key must be below the bottom pins in order to put the key into the lock. The distance from the top of the blade of the key to that point is how much space the manufacturer has to use for possible cut depths. Note that both of the following keys have ten possible depths, but because the tip is lower on one, the size of the increments can be larger.

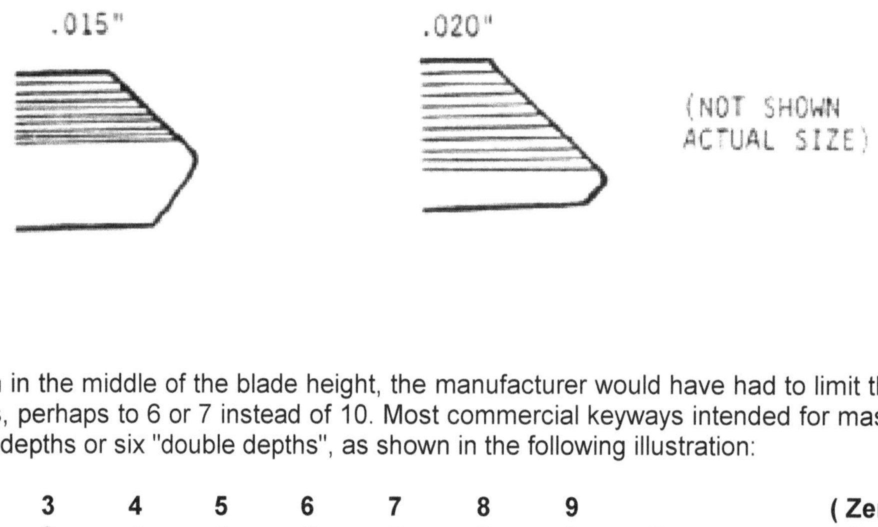

If the tip had been in the middle of the blade height, the manufacturer would have had to limit the number of possible depths, perhaps to 6 or 7 instead of 10. Most commercial keyways intended for master keying have ten possible depths or six "double depths", as shown in the following illustration:

0	**1**	**2**	**3**	**4**	**5**	**6**	**7**	**8**	**9**			**(Zero to 9)**
	1	**2**	**3**	**4**	**5**	**6**	**7**	**8**	**9**	**0**		**(One to Zero)**
	1		**2**		**3**		**4**		**5**		**6**	**(1 to 6)**
0	**9**	**8**	**7**	**6**	**5**	**4**	**3**	**2**	**1**			**(Zero to One)**

The importance of a larger increment is to prevent **INTERCHANGE**, which is when a key operates a lock cylinder or core that it was not intended to operate due to tolerances within the cylinder.

When we say **INCREMENTS**, we refer to the standardized possible differences in depth chosen by the manufacturer.

When we go from one possible depth in a given position to the next possible depth for that position, we have moved "one increment". The key as it was is now deeper in one position by one increment (also called one "step").

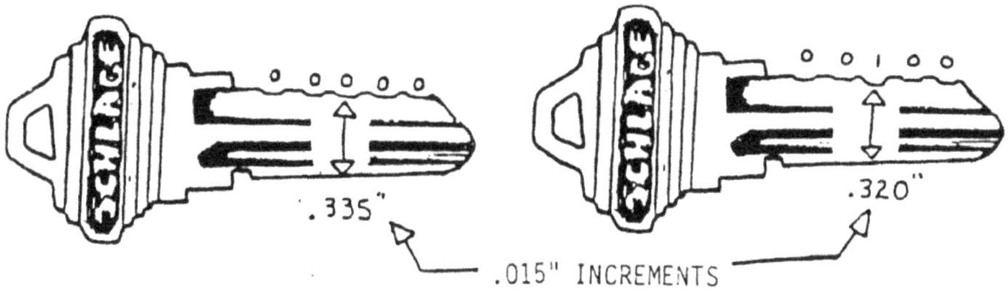

If we do it in two positions, it is still only one increment deeper anywhere on the key, but it is one increment deeper in two positions (or spaces).

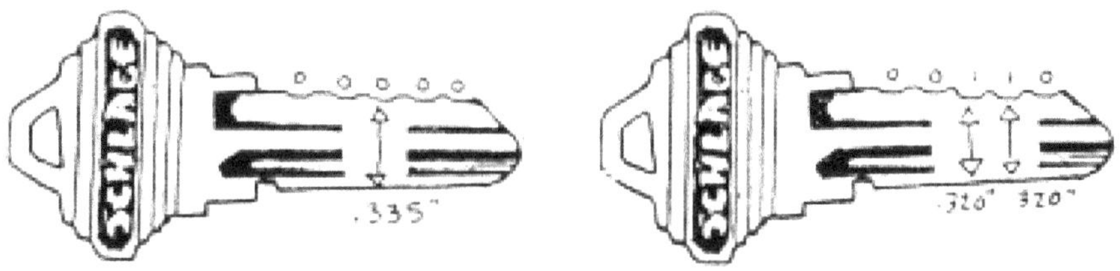

In our example above, the size of the increment is .015" (fifteen thousandths of an inch), which means that we could have said it was "fifteen thousandths of an inch deeper in two positions". But "one step" is less to say than "one increment", which is less than "fifteen thousandths of an inch".

Some key blanks are not a proper fit in the keyway, permitting the key to be rocked up and down by as much as .018" (eighteen thousandths of an inch). If the increments of that lock were .018" or less, a one increment difference would not be sufficient to keep the wrong key from operating the lock cylinder.

One cylinder on the market uses a flat at the top of the plug that creates up to a .023" (twenty three thousandths of an inch) gap. Once again, a one increment difference would not be sufficient to prevent interchange even with .020" (twenty thousandths of an inch) increments!

Original pins from the manufacturer are usually numbered to match the depth increments on the key, so a number 0 cut takes a number 0 bottom pin; a number 1 cut takes a number 1 bottom pin; a number 2 cut takes a number 2 bottom pin; etc.

Replacement pin kits are numbered in thousandths of an inch, based on the length of the pin. The differences in increments of length in these pin kits may be .0025" (two and a half thousandths), .003" (three thousandths), .005" (five thousandths), or .010" (ten thousandths). A ten thousandths (.010") pin kit is definitely NOT recommended for master keying purposes. Any of the others will suffice, with varying degrees of efficiency depending primarily on the manufacturer whose brand of locks you are working on, and the tolerances, wear factors, etc., in them.

A pin kit in the manufacturer's exact pin lengths is, of course, IDEAL for master keying purposes, but this is not always possible for a wide variety of reasons including availability, cost factors, and time limitations.

There is some basic math which is applied to pin length selection. It is:

EFFECTIVE SHEARLINE= BOTTOM PIN LENGTH
\+ KEY CUT "DEPTH"

Or, expressing it differently,

BOTTOM PIN LENGTH = EFFECTIVE SHEARLINE
\- KEY CUT "DEPTH"

Or, differently again,

KEY CUT = EFFECTIVE SHEARLINE
\- BOTTOM PIN LENGTH

This is because, ideally, the key sits with the bottom of the key at the bottom of the plug, and the bottom pin sits on the key at the key cut, with its top at the shearline at the top of the plug.

We said "IDEALLY" because there are several other factors to consider. First of all, the plug is round. In order for this to be entirely accurate, the bottom of the key would have to be fully radiused to match the bottom of the plug. Some original keys and key blanks are radiused this way. Others have the edges shaved or chamfered. Most replacement key blanks are flat on the bottom.

The fully radiused blank seats fully in the plug. The shaved edge blank leaves a gap of approximately .002", and the flat bottomed key sits approximately .005" higher than the bottom of the plug.

This means that a key cut by code on a flat bottom key blank, to operate the same bottom pin length, must be cut .005" deeper. A shaved edge key blank must be cut .002" deeper. Problems can also occur during duplication attempts.

The other factor involved is the bottom pin. In order to fully fit the top of the plug, the pin would have to be radiused to the same curvature as the plug also. Most bottom pin tumblers, however, have only a slight chamfer at the edges. Thus the edges would still bind if we used the outside plug diameter for our math. Using the chamfered edge bottom pin, the EFFECTIVE shearline is actually about .004" lower than the outside plug diameter.

Thus, with an outside diameter of .509", and using flat bottomed key blanks and chamfered edge bottom pins, our effective shearline for the pinning math would be .500".

The .500 is calculated as follows:

.509" Plug diameter
- .004" To compensate for the chamfer
= .505"
- .005" To compensate for the flat key blank
= .500" Effective shearline

If we had been using key blanks that were shaved on the edge, the math would have been calculated as follows:

.509" Plug diameter
- .004" To compensate for the chamfer
= .505"
- .002" To compensate for the shaved edge blank
= .503" Effective shearline

This is more math than most locksmiths want to deal with and can be totally avoided if using manufacturer's original keys and manufacturer's original pins and springs.

Even for a particular manufacturer, the tolerances within the cylinder may vary with the type of cylinder or the materials from which it is made.

The most common types of cylinders are shown in the next illustration:

Rim

U.S. Mortise

An example of one kind of Key in Knob cylinder

Key in Leverset

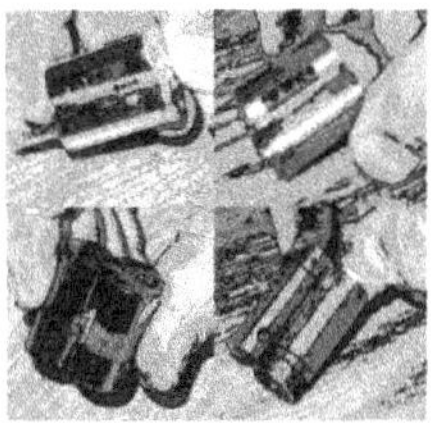
interchangeable and removable cores

Residential Key in Knob Cylinder

Regardless of the type of cylinder, it will have a number of chambers (holes) which match the spacing set up by the manufacturer and into which are placed the bottom pin tumbler (which has a round or conical end or ends), any master pins, build-up pins or control pins (Don't worry- we will cover these shortly) and a driver pin, topped by a spring of appropriate length to fit without being crushed.

All of these will be held in place, usually by some sort of removable retainer cap or pressed-in slide, which would have to be removed in order to service the pin tumblers from the top. (This is NOT the only way.)

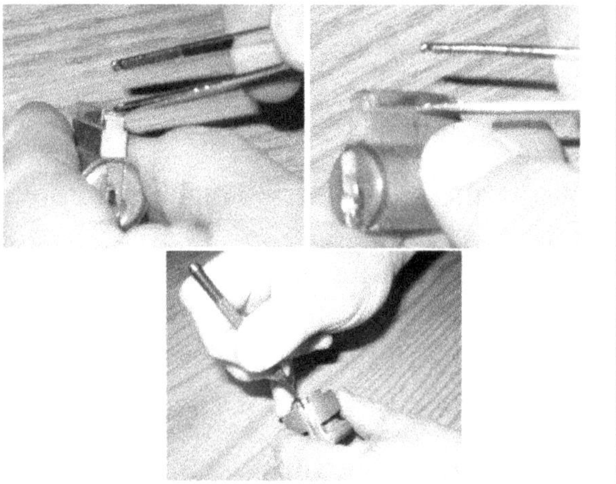

Another way is to either use the key or use a piece of SHIM STOCK (a small, very thin sheet of metal that can fit into the cylinder and be used to separate the tumblers).

First, remove the retainer on the back of the plug.

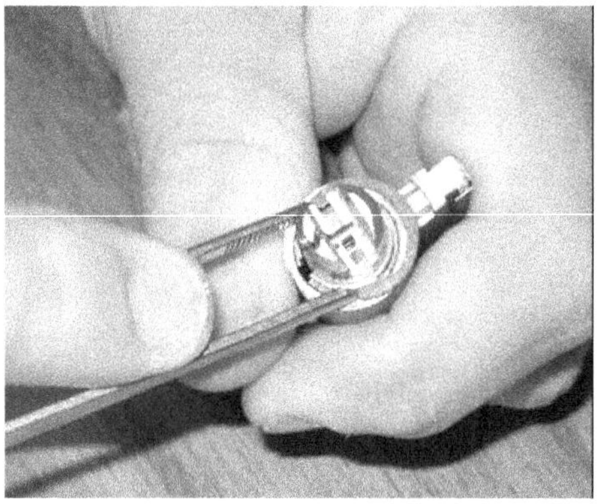

Then use a key, pick the cylinder, or shim it using shim stock and a blank.

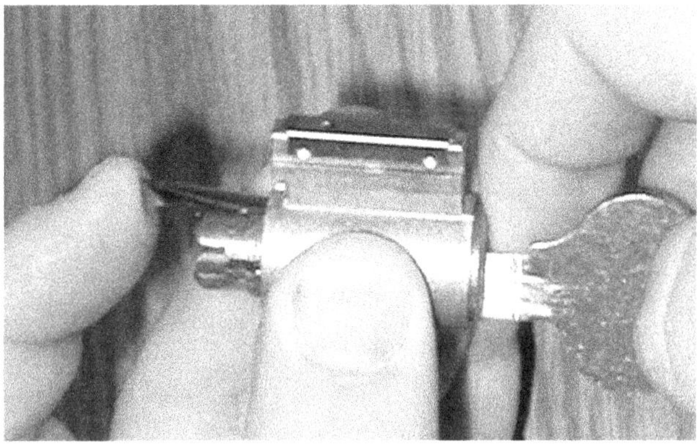

Once you have the plug turned slightly you are ready to use a **PLUG FOLLOWER.**

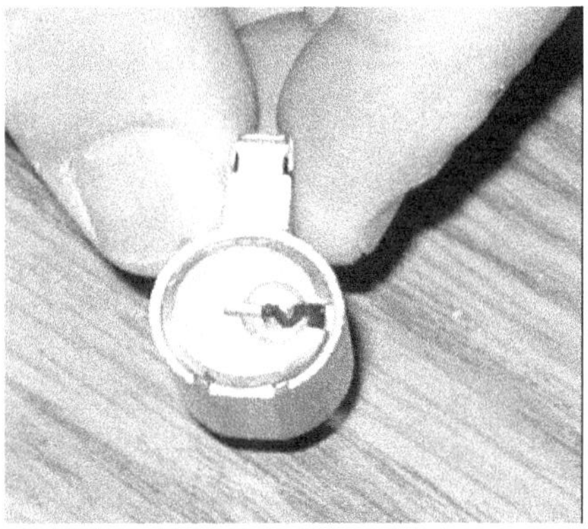

Rotate the plug to 90 degrees from the chamber location.

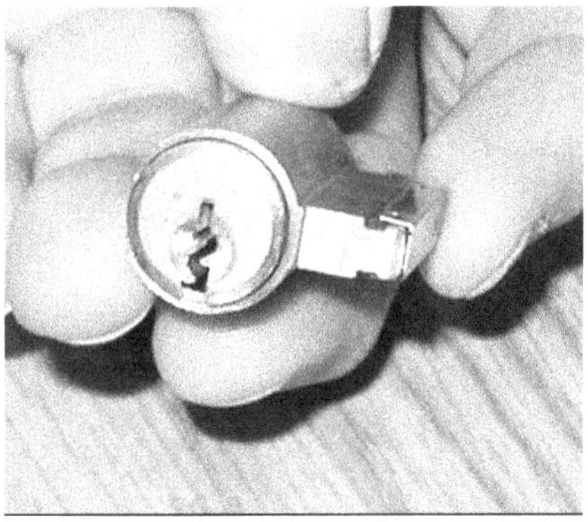

Then rotate the cylinder so that the top of the keyway is "up".

Hold slight pressure against the keyway and apply pressure to the plug follower.

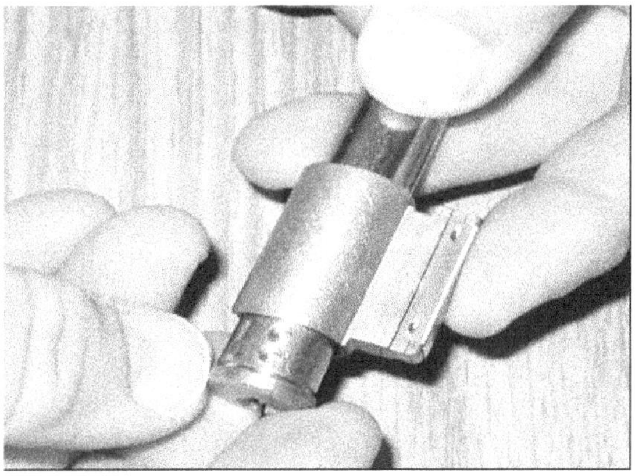

Maintain the pressure to keep plug and plug follower together.

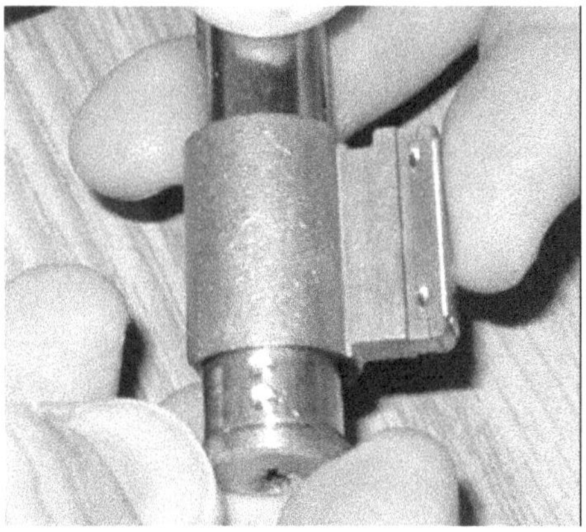

Remember to keep the keyway "up" so the bottom pins do not spill.

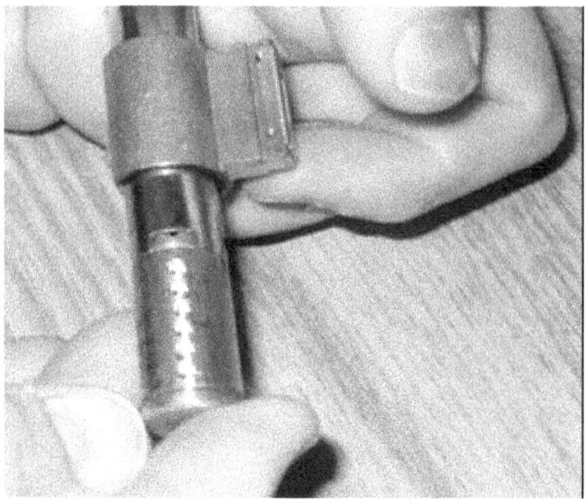

When the plug is fully out set both parts where they will not spill.

In a master keyed cylinder, a smaller pin, usually flat on both top and bottom, makes up the difference between the two cuts that need to operate a position. This shows a #3 cut and a #5 cut both operating on opposite ends of the master pin.

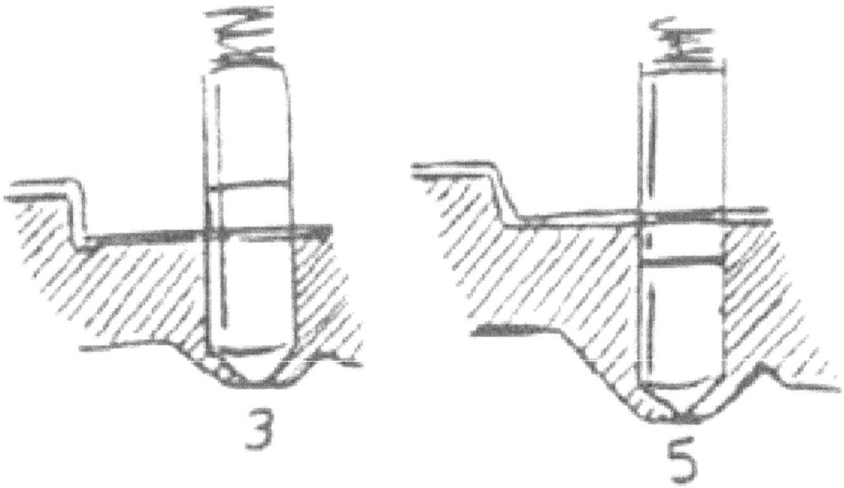

Bottom pins will match the smallest of the two cut depths, in this case a #3, and the master pin will make up the difference between the two lengths, which in this case is a #2 master pin.

A typical universal pin kit shows the possibilities, based on length of the pin:

Key Cut	Bottom Pin	Master Pin
0	.165	
1	.180	
2	.195	.030
3	.210	.045
4	.225	.060
5	.240	.075
6	.255	.090
7	.270	.105
8	.285	.120
9	.300	.135

Note that there is no master pin shown in the previous chart smaller than a #2. That is because the #1 for this lock would be .015", which is too thin to be effectively and safely used in ANY existing cylinder on the market today.

To determine the pinning for a master keyed cylinder, determine the smallest pin required between the two keys for a specific position, and use the bottom pin to match it. Then choose a master pin to match the DIFFERENCE between the two cuts for that position. Do that for each position on the two keys (top master key and operating (or change) key.

Key One:	3	4	2	6	3	7
Key Two:	5	2	4	2	5	9

Difference:	2	2	2	4	2	2
Shortest:	3	2	2	2	3	7

The Difference is shown on top, as this is how the master pins sit in relation to the bottom pins. Older charts may show this in a different pattern, so be careful when reading pre-existing charts!

Steps To Master Pin Cylinders

- **Calculate Bottom Pins**
- **Calculate Master Pins**
- **Insert Bottom Pins Into The Plug**
- **Insert Master Pins On Top Of Bottom Pins**
- **Re-Inserted The Plug Into The Shell**
- **Fasten The Plug Retainer in Place**
- **Try All Keys**

If the top pins (drivers) need replaced (always a necessity with used or existing cylinders rather than new ones) insert the spring using a pair of tweezers. There are many types. Find one that is comfortable for you for each task. Then, using the tweezers, pick up the driver with most of it exposed. Use it to push down the spring and use the plug follower to hold it in place. Finally push it the rest of the way down, permitting the plug follower to hold it in.

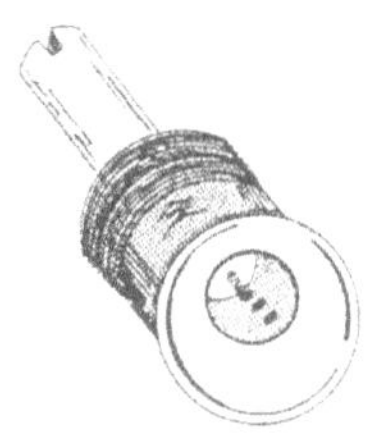

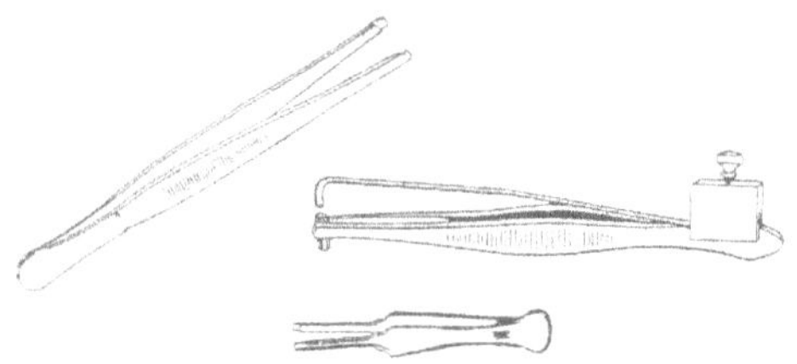

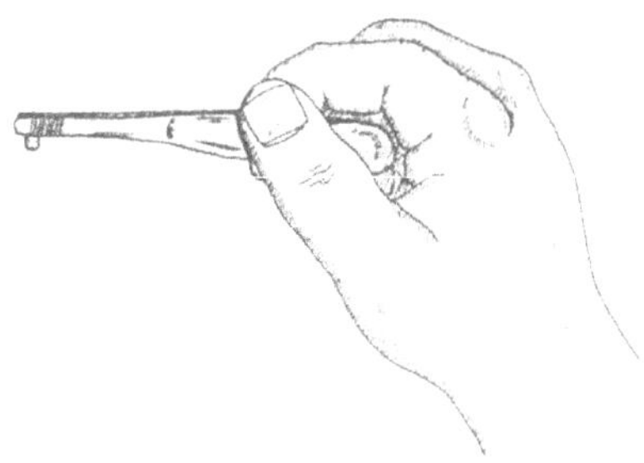

Note the major portion of the driver protruding from the base of the tweezers.

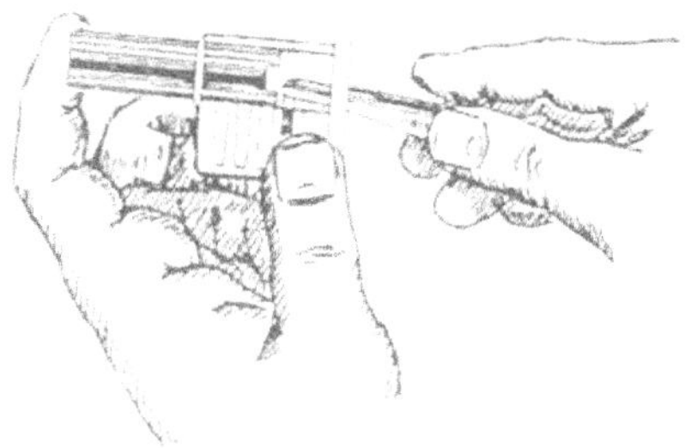

"X-Ray" views of the interaction between the follower, tweezers and driver pin.

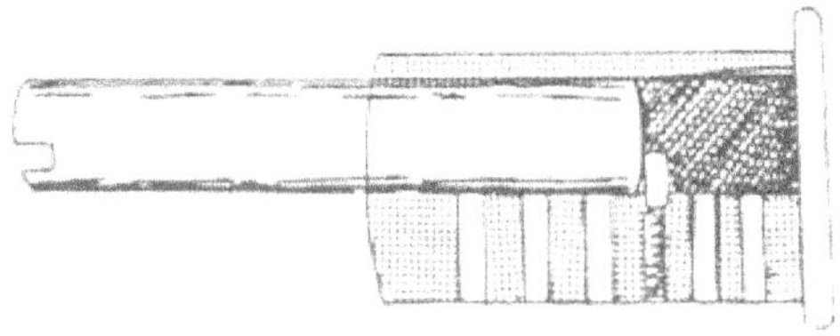

Generally speaking, the EXACT driver size is not all that critical in most lock cylinders. In fact, even in cylinders with an exact matched driver size, a difference of a few thousandths of an inch will rarely, if ever, affect the working of the cylinder. Nevertheless, if we use the drivers used by the manufacturer and by other locksmiths successfully, we have reduced the likelihood of ever having a problem with a cylinder.

The following is a composite of the driver sizes listed as appropriate from manufacturers, or found by measuring drivers from existing working cylinders:

MANUFACTURER	DEPTHS	DRIVER SIZE
ARROW	ALL except SFIC	.155
CORBIN	ALL (except IC)	.172
FALCON (REGULAR)	ALL (except IC)	.186
KWIKSET	ALL	.180
LOCKWOOD	ALL (except IC)	.220
RUSSWIN	ALL (except IC)	.172
SARGENT	1-3	.220
	4-6	.180
	7-0	.140
SCHLAGE	0-3	.235
	4-6	.200
	7-9	.165
SEGAL	ALL	.174
WEISER	ALL	.186
WESLOCK	0-3	.170
	4-9	.126
YALE	0-2	.231
	3-5	.190
	6-8	.150
	9	.114
YALE ALSO LISTED AS:	0-3	.260
	4-6	.230
	7-9	.190

Some locksets, as you may have noticed in the driver length chart, use varying sizes of drivers to match the length of the bottom pin, with a long driver for short bottom pins, and a short driver for long bottom pins. This is referred to as **COMPENSATED DRIVERS**. **You may occasionally hear it referred to as Matched Pins.**

Related Terms from the LIST Council Dictionary:

compensate drivers
v. to select longer or shorter top pins, depending on the length of the rest of the pin stack, in order to achieve a uniform pin stack height

Compensated Drivers tend to be preferred for a number of reasons, including reduced wear, increased pick resistance, and cylinder integrity.

These can be roughly matched into several sizes, or they can be exactly matched with an "**EXACT STACK HEIGHT**". For example, the chart shows Schlage using a .235" (long) driver with bottom pins of 0,1,2,3. It uses a .200" (medium) driver with a 4,5,or 6 bottom pin. And it uses a .165" (short) driver with a 7,8,or 9 (long) bottom pin. This is commonly referred to as "**SIZED DRIVERS**" or "**MATCHED PINS WITHOUT AN EXACT STACK HEIGHT**". Note that these terms are not yet popular enough to be included in the LIST Dictionary, so you should be careful with their usage.

Commercial Medeco, on the other hand (not shown on the chart), uses an exact stack height. Each bottom pin has an exact length of driver to match it, and if you were to add the bottom pin length and the driver length together, it would add up to an exact figure. In Medeco's case (since 1977), that figure is .506". Thus the following would apply:

COMMERCIAL MEDECO

A BITTING OF	USES BOTTOM PIN	USES DRIVER
1	.236	.270
2	.266	.240
3	.296	.210
4	.326	.180
5	.356	.150
6	.386	.120

NOTICE THAT EACH NUMBER BOTTOM PIN ADDED TO ITS DRIVER=.506"

Any time we are master keying a cylinder which uses matched pins (whether "sized" or with an exact "stack height") we must work our pinning math slightly differently to compensate for it.

STANDARD PINNING MATH FOR A TYPICAL CYLINDER WOULD BE:

MASTER KEY BITTING: 321414
CHANGE KEY BITTING: 143232

SMALLEST # IN PAIR = BOTTOM PIN 121212
DIFFERENCE IN PAIR = MASTER PIN 222222

BUT NOW WE HAVE TO ALSO CHOOSE THE NUMBER TO USE FOR EACH DRIVER. THE RULE IS AS FOLLOWS:

LARGEST IN PAIR OF NUMBERS PER CHAMBER = DRIVER SIZE

SO OUR MATH, INSTEAD, LOOKS LIKE THIS:

HIGHEST MASTER KEY ACTUALLY USED:	**321414**
CHANGE KEY BITTING:	**143232**
SMALLEST IN PAIR = BOTTOM PIN	**121212**
DIFFERENCE BETWEEN PAIR = MASTER PIN	**222222**
LARGEST IN PAIR = DRIVER	**343434**

The reason for using the largest number in each pair is simple. The space taken up by the bottom pin and master pin in the chamber will be equal to the length of the longest bottom pin that would have been required by either key. Thus the driver size has to be of the size it would call for.

This also allows for a double check of the math, because after we have calculated the math as described, we can add the bottom pin and master pin in each position together, and they should equal the driver length we chose. Once the drivers and springs are inserted and the lock reassembled (so it does not fall apart when you try the keys), you are ready to try the keys! If they are not already cut, you will have to cut them, **BY CODE**!!!

USE OF DEPTH KEY SETS IS NOT CONSIDERED GEBERALLY CONSIDERED A PROFESSIONAL TECHNIQUE FOR THE PURPOSES OF MASTER KEYING!!! IN ORDER TO GUARANTEE NO INTERCHANGE, THE KEYS MUST BE CUT EXTREMELY ACCURATELY. DEPTH KEY SETS AND DUPLICATING MACHINES DO NOT GENERALLY OFFER THIS TYPE OF ACCURACY!!!

If you must use depth keys or a partial code machine, take your time to be as accurate as possible (and put at least half the profits away to buy a code machine).

Lesson Four:
PREPARING FOR PROGRESSION

In a master key system, all the keys are related to each other, and to the master key. This relationship allows the system to work, with a specific job assigned to each key.

These relationships all begin with the highest level of master key in the system. This is because there are many ways that the change keys can be related to each other, but there are only three ways that they can be related to the highest master key:

POSSIBILITY ONE: **Development Rule: None of the change keys has the same depth of cut in the same position as any of the cuts on the highest master key.**

EXAMPLE: MASTER 35241

53423	53623	57423	57623
53425	53625	57425	57625
53427	53627	57427	57627
53463	53663	57463	57663
53465	53665	57465	57665
53467	53667	57467	57667

Note that none of the change keys has a #3 in the first position, a #5 in the second position, a #2 in the third position, a #4 in the fourth position, or a #1 in the fifth position.

This possibility deals with the method of master keying known as **TOTAL POSITION PROGRESSION**". Other terms you may see in older texts include **"ACROSS THE KEY PROGRESSION**", and" **SIMPLE PROGRESSION".**

Related Terms from the LIST Council Dictionary:

total position progression
n. a process used to obtain key bittings in a master key system wherein bittings of change keys differ from those of the top master key in all bitting positions

POSSIBILITY TWO: **Development Rule: The change keys have certain positions where ALL of them have the same depth of cut as on the highest master key.**

EXAMPLE: MASTER 35241

35423	35623	35823
35425	35625	35825
35427	35627	35827
35463	35663	35863
35465	35665	35865
35467	35667	35867

Note that all the keys have a #3 and a #5 in the first two positions respectively, but that none of them has a #2 in the third or a #4 in the fourth, or a #1 in the fifth.

This possibility deals with a specialized technique for master keying known as **LIMITED PROGRESSION.** (In this case, it would be a **THREE COLUMN PROGRESSION**, since there are three positions where the change keys are different from the highest master key.) This technique is actually a subset of the RCM (Rotating Constant Method) but is sometimes used alone, especially by commercial master keying locksmiths

POSSIBILITY THREE: **Development Rule: Each of the change keys has a certain NUMBER of cut depths in common with the highest master key, although exactly WHICH cuts they are varies from change key to change key.**

EXAMPLE: MASTER 35241

35243	35221	35441	31241	15241
35245	35261	35641	33241	55241
35247	35281	35841	37241	75241

Note that all of the change keys in this Example have exactly four cuts in common with the highest master key, and one position which is different from it, but that it is not the same position on all the keys.

In the first group, the far right position is the only one that is different; in the second group it is the fourth position from the left; in the third group it is the middle position; in the fourth group it is the second from the left; and in the fifth group, the far left position is the one which differs.

This possibility deals with the method of master keying known as **ROTATING CONSTANT METHOD**. It was once also known as **HOLD & VARY** but this term has fallen from public acceptance.

Related Terms from the LIST Council Dictionary:

rotating constant
n. one or more cut(s) in a key of any level which remain constant throughout all levels and are identical to the top master key cuts in their corresponding positions. The positions where the top master key cuts are held constant may be moved, always in a logical sequence

rotating constant method
n. a method used to progress key bittings in a master key system, wherein at least one cut in each key is identical to the corresponding cut in the top master key. The identical cut(s) is moved to different locations in a logical sequence until each possible planned position has been used.

In addition to the three change key to master key relationships shown above, there is also another possible relationship known as **POSITIONAL MASTER KEYING**. It consists of leaving certain positions completely without tumblers and varying which positions they are. This, however, is not especially suited for use in standard pin tumbler locks, and is usually reserved for special mechanisms, such as KABA and DOM high-security multiple cylinders and SCHLAGE Wafer locks (as well as similar products).

Related Terms from the LIST Council Dictionary:

positional master keying
n. a method of master keying typical of certain binary type disc tumbler key-in-knob looks and of magnetic and dimple key cylinders. Of all possible tumbler positions within a cylinder, only a limited number contain active tumblers. The locations of these active tumblers are rotated among all possible positions to generate key changes. Higher level keys must have more cuts or magnets than lower level keys.

There is also a specialized technique that looks like "no system" but actually has one, recently named as **SPHERICAL MASTER KEYING**, but that is not covered in this book because although the technique has existed since the late 1970's it is still in its infancy at the time of this writing.

By the way, a "system" that has not been developed properly but is simply a random grouping of keys is generally referred to by the trade as a "shoebox" system, based on a perception that unskilled locksmiths used to prepare master key systems by pulling keys at random out of a shoebox.

It is not a good idea to use the above to try to "crack" a master key system or to "key into an existing master key system.". If you have ALL the keys that are in use, and know which keys have ALREADY been changed, the above would work. Otherwise you put the system and yourself at risk. Although the rules describe the conditions by which a system is to be progressed, unless you have access to ALL the key bittings in a system, you cannot guarantee that the system was developed properly following the rule throughout in the past.

In fact, this is so important that the ALOA (Associated Locksmiths of America) put it into their Technical Standards Policy (ver. 2.3, 05/01/2005) the following:

11. Master Key System Integrity: No attempt will be made to expand any master keyed system without first obtaining a valid key bitting array and list of all key bittings currently in use. If this is not possible, the lock shall be keyed SKD or a new master key system generated.

In master keying we will be making sure that the change keys differ from each other by enough that they should not be capable of operating each other's locks. That may be a one increment difference, a two increment difference, or may even require a three increment difference.

What decides how much difference is required?

1) Size of the increments
2) Tolerances within the lock
3) The number of change keys required by the system

Let's look at these three items one at a time.

SIZE OF INCREMENTS: Some key blanks aren't a tight fit in the keyway, and the bow of the key can be rocked up and down by as much as .018". If the increments of that particular lock were .018" or less, then a one increment difference could be comprised by "rocking" the key. In that particular lock, a two increment or larger difference should be used.

TOLERANCES WITHIN THE LOCK: At least one lock cylinder on the market today has a "FLAT" on the top of the plug, that creates a .023" gap between the plug and shell at the shear line In that particular cylinder, which uses .018" increments, a MINIMUM of a two increment difference is required, and a three increment difference might be a much better idea, especially when you consider the chances of "rocking" the key as well. Also, any keyway, when worn, will allow the key to be "rocked" to a greater extent than in a new cylinder.

THE NUMBER OF CHANGE KEYS REQUIRED: Occasionally the system will be so large that the number of increments suggested by the two previous guidelines will not allow as many different change keys as are required by the customer. In this case, a better idea might be to replace the locks with ones with fewer limitations. However, the customer may be less concerned with "key wiggling" than with the cost of replacing the locks. In this case, you may choose to use a smaller number of increments of difference than you might otherwise have chosen. (The smaller the increments of difference, the more change keys the system will yield. This can be very important in locksets where the manufacturer's numbering system has a smaller number of available keys to begin with. More on this will be given later in this lesson, and will be discussed in depth in later lessons.)

You have just been introduced to the four potential "rules" by which master key systems can be progressed—Total Position Progression (formerly Simple Progression or Across the Key Progression), Limited Progression (not an actual method but a potentially useful technique), Rotating Constant (formerly Hold & Vary), and Positional Master keying, as well as to the various step options which can be used with each--a single increment of difference between change keys, a two increment difference between change keys, a three increment difference between change keys, or some combination of the various possible differences between change keys, with one of the differences in one or more positions, and a different amount of differences in other positions.

But now it is time to meet one of the most valuable tools in master keying. It is a small chart called the **KEY BITTING ARRAY** or **KBA**.

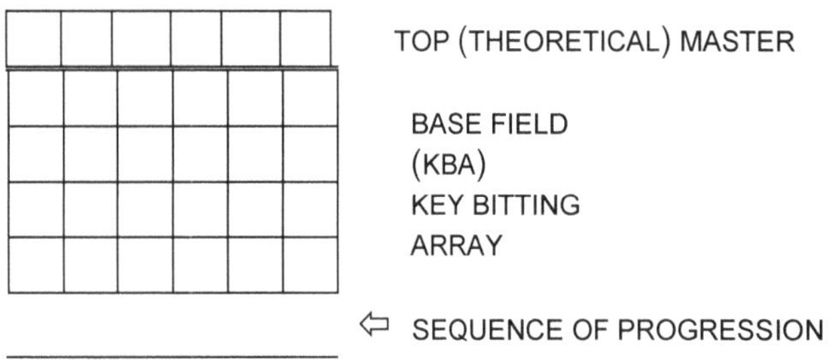

Over the years, this chart had its face lifted and its name changed a number of times. It was known as **The Progression Formula Chart** and the **Base Field Development Chart.** These terms have almost completely fallen out of popularity. They are referred to here only to allow persons who have previously learned the old terms from older texts to make the connection to the new term.

Related Terms from the LIST Council Dictionary:

key bitting array
n. a matrix (graphic) display of all possible bittings for change keys and master keys as related to the top master key

KBA
abb. key bitting array

This chart is probably the single most important tool in all of master keying, since we use it to determine everything else that we will be doing to select the keys for the system.

As you can see, the chart is divided into three sections.

The top section, which has five, six or seven spaces across to be filled with numbers, is for us to list our **Top Master Key** bitting which we will use to develop all the other areas of the full master key chart.

> **Related Terms from the LIST Council Dictionary:**
>
> **top master key**
> n. the highest level master key in a master key system

The middle section, which we will fill with numbers representing the possible cuts reserved for change keys, is the **POSSIBLE DEPTHS** section (actually no one has ever specifically named this section, but that is what fills it).

The bottom section is for us to describe the order in which the positions on the key will be progressed, and is known as the **SEQUENCE OF PROGRESSION** or **SOP**. Letters in lower case (a not A) will be used to decide which position gets progressed first, second, third, etc. At one time people used numbers there, but this caused errors to appear in the charts as these numbers were occasionally brought down into the chart by error as though they were change key bittings.

> **Related Terms from the LIST Council Dictionary:**
>
> **sequence of progression**
> n. the order in which bitting positions are progressed to obtain change key combinations

The following shows the numbering systems for some locksets on the market in the U.S.

MANUFACTURER & SERIES	INCREMENTS	NUMBERING SYSTEM
Arrow (not SFIC)	.018"	0 to 9
Arrow (SFIC) Small Format Interchangeable Core (A2)	.0125"	0 to 9
Arrow (SFIC) Small Format Interchangeable Core (A4)	.0175"	0 to 5
Best (SFIC) Small Format Interchangeable Core (A2)	.0125"	0 to 9
Best (SFIC) Small Format Interchangeable Core (A3)	.0175"	0 to 6
Best (SFIC) Small Format Interchangeable Core (A4)	.0205"	0 to 5
Corbin S	.014"	1 to 0
Corbin X pre 1970	.014"	1 to 0
Corbin X System 70	.028"	1 to 6
Corbin Z. (1969)	.014"	0 to 9
Corbin-Russwin System 70	.028"	1 to 6
High Security (Corbin-Russwin) Emhart Interlocking Pin	.028"	1 to 6
Falcon Interchangeable Core (A2)	.0125"	0 to 9
GL (SFIC) Small Format Interchangeable Core (A2)		0 to 9
Ilco (KABA-ILCO)	.018"	0 to 9
Kwikset	.023"	1 to 7 ##
KSP (SFIC) Small Format Interchangeable Core (A2)	.0125"	0 to 9
Lockwood Multiplex	.018"	1 to 0
Lockwood (SFIC) Small Format Interchangeable Core (A2)	.0125"	0 to 9
Medeco (Sidebar Large Pin)	.030"	1 to 6
Medeco (Restricted Large Pin)	.015"	0 to X (11 d)
Russwin Multiplex (D,H,)	.015"	0 to 9
Russwin 852B	.020"	0 to 9
Russwin System 70	.028"	1 to 6
Sargent	.020"	1 to 0
Schlage (Including Everest C and D, Primus, FSIC, Classic)	.015"	0 to 9
Schlage (SFIC) Small Format Interchangeable Core, including Everest B (A2)	.0125"	0 to 9
Weiser	.018"	0 to 9
Weslock	.0155"	0 to 9
Yale Multiplex & Grade One 8 keyway	.019"	0 to 9 ***

The #7 cut does not usually appear on factory keys for individual locks.

*** Yale residential grade two is usually .025"

It is accepted in the trade that it is generally necessary to have a minimum of .023" difference in average tolerance cylinders to prevent interchange, although the .020" used by A4 cores is also generally considered acceptable.

For that reason, all of the manufacturers above with either a 0 to 9 or 1 to 0 numbering system would require a minimum of two increments between their cuts and any different cuts in the same position on other keys to prevent interchange in a typical cylinder or core.

All of the manufacturers above with a 1 to 6 or 0 to 5 can typically accept a single increment of difference and still generally prevent interchange in a typical cylinder or core.

Once you know the numbering system your lock manufacturer has chosen, you are ready to pick the top (theoretical) master key cuts and fill out the rest of the KBA.

A little history might be useful here. Wiegend Jensen and Bert Michaels had a unique way of introducing the **Key Bitting Array (KBA)** that makes it more obvious what its function is and where the numbers to fill it out come from.

They started out with a slightly different looking chart that had five, six or seven positions across (to represent the positions on the key in a five, six or seven pin cylinder, just like the standard KBA of today. However, instead of having four or five rows beneath the **Top (theoretical) Master Key (TMK)** and above the **Sequence of Progression (SOP),** they had eleven rows, populated with the numbers 0 (shallow) to 0 (deep), as shown here:

0	0	0	0	0	0
1	1	1	1	1	1
2	2	2	2	2	2
3	3	3	3	3	3
4	4	4	4	4	4
5	5	5	5	5	5
6	6	6	6	6	6
7	7	7	7	7	7
8	8	8	8	8	8
9	9	9	9	9	9
0	0	0	0	0	0

These numbers represented all the possible numbering systems in a six pin standard cylinder. The next thing that needed done was to cross off any that did not apply to the master key system we are developing.

The first step in filling out their chart was to cross off any spaces not being used to represent positions for cuts on the key, so if this were a five pin cylinder, we would cross off one column (on the left or right, whichever makes more sense to you. After all, this would be your tool.

For now, we will assume we are working with a six pin cylinder, so the six spaces across each row are just fine as they are.

Now we have to determine which cuts do not match our manufacturer's numbering system and cross them off.

If the numbering system were 0 to 9, the extra zeros across the bottom would be crossed out, as shown here:

0	0	0	0	0	0
1	1	1	1	1	1
2	2	2	2	2	2
3	3	3	3	3	3
4	4	4	4	4	4
5	5	5	5	5	5
6	6	6	6	6	6
7	7	7	7	7	7
8	8	8	8	8	8
9	9	9	9	9	9
0	0	0	0	0	0

If instead we were using a manufacturer whose cut depths were 1 to 0, we would cross off the row of zeros at the top instead, as shown:

0	0	0	0	0	0
1	1	1	1	1	1
2	2	2	2	2	2
3	3	3	3	3	3
4	4	4	4	4	4
5	5	5	5	5	5
6	6	6	6	6	6
7	7	7	7	7	7
8	8	8	8	8	8
9	9	9	9	9	9
0	0	0	0	0	0

If it were 1 to 6:

0	0	0	0	0	0
1	1	1	1	1	1
2	2	2	2	2	2
3	3	3	3	3	3
4	4	4	4	4	4
5	5	5	5	5	5
6	6	6	6	6	6
7	7	7	7	7	7
8	8	8	8	8	8
9	9	9	9	9	9
0	0	0	0	0	0

and for 0 to 5 it would end up like this:

0	0	0	0	0	0
1	1	1	1	1	1
2	2	2	2	2	2
3	3	3	3	3	3
4	4	4	4	4	4
5	5	5	5	5	5
6	6	6	6	6	6
7	7	7	7	7	7
8	8	8	8	8	8

Once we had the chart developed to this stage, it is time to make three decisions--the highest master key bitting, the number of increments of difference we decided was necessary, and the master keying layout rules we will apply to the system.

Let's start with the step options, or the number of increments difference our system will need.

As we saw just below the manufacturer's numbering system charts, our MINIMUM is either a single increment or two increments. These can also be referred to as a **SINGLE** STEP or a **TWO STEP**.

They are the basic choices, and usually the only ones they will encounter. However, to compensate for wear or tolerances, it may be necessary to choose a different option.

SINGLE INCREMENT or **SINGLE STEP -** a single increment of difference will be sufficient between master key and change keys, and between change keys. This option does not affect the TMK selection.

MASTER	548210
CHANGE KEYS	657121
	657122

LIMITED ONE STEP - two increments of difference between the master and any change keys is used, but only a one increment difference between the change keys

MASTER	548210
CHANGE KEYS	326432
	326433

The numbering system affects the possible master key choices. In a Zero to Nine numbering system, the use of Zero or Nine cuts on the master key increases the number of possible combinations for change keys. In a One to Zero, the use of a One or Zero cut on the master key has the same effect. All other master key bittings for either of those systems are equal. In a Zero to Six numbering system, the use of Zero or Six depths on the master key increases the number of possible change keys. In a One to Seven numbering system, the use of a One or Seven depth will give the most theoretical key changes. In a Zero to Eight numbering system, the use of Zero or Eight depths on the master key increases the number of theoretical change keys available to us for our system. In a numbering system of One to Nine, the use of One or Nine depths increases the number of potential change keys. In a One to Six numbering system, it is the use of One or Six depths that increases the number of change keys.

To sum it all up, in this format, the use of the shallowest and deepest of the possible depths available in the numbering system will yield the maximum number of possible change keys when using this step option.

TWO INCREMENT DROP or **TWO STEP -** two increments of difference is used between the highest master key in the system and any of the change keys, as well as between any of the change keys and other change keys.

MASTER	548210
CHANGE KEYS	326032
	326034

This results in a **PARITY PATTERN** of Odd and Even positions, based on the TMK (Top Master Key) selected.

Related Terms from the LIST Council Dictionary:

parity
n. the property of being either even or odd, as the depth number of a bitting in a combination, or the possible progressives in a given position of a two step progression

parity pattern
n. the collective description of the parities of a group of bitting positions in a two step progression, typically expressed in an even/odd sequence, e.g. EOOEEO

Some patterns yield more unusable or undesirable keys than other patterns. For cylinders with a 0 to nine or one to zero numbering system, the 26 optimal patterns are:

OOEOEE	**OOEEOO**	**OOEEOE**
EEOEEO	**EOEEOE**	**EOOEEO**
EEOEOE	**EOEEOO**	**EOOEOE**
EEOEOO	**EOEOEE**	**EOOEOO**
EEOOEE	**EOEOEO**	**OEEOEE**
EEOOEO	**EOEOOE**	**OEEOEO**
OEOOEO	**OEOEEO**	**OEOEOO**
OEEOOE	**OEOEOE**	**OEOOEE**
OOEOEO	**OOEOOE**	

In Zero to Six or a Zero to Eight numbering system, the pattern of odds and evens is even more important, because the more even numbered cuts on the highest master key, the more change key combinations are theoretically possible.

The opposite is true on a One to Seven or a One to Nine numbering system, where the more odd numbered cuts on the highest master key, the more change keys are theoretically available for use in developing the master key system.

The two increment drop is rarely used in locks with a numbering system of One to Six, but if it were, the 26 acceptable patterns listed above would be the only new guideline.

THREE INCREMENT DROP or THREE STEP - three increments of difference between the master key and any of the change keys, as well as between any of the change keys and other change keys.

MASTER	<u>548210</u>
CHANGE KEYS	215543
	215546

With this step option, there is relatively no difference between any of the acceptable key bittings for the highest master key.

However, to maximize the number of change keys, the following will make a slight increase when used on master keys:

NUMBERING SYSTEM	OPTIMUM CUT DEPTHS
0 to 9	0,3,6,9
1 to 0	1,4,7,0
0 to 6	0,3,6
1 to 7	1,4,7
0 to 8	ALL EQUAL
1 to 9	ALL EQUAL
1 to 6	ALL EQUAL

LIMITED THREE STEP - a three increment difference between the highest master key and any of the change keys, but only a two increment difference between change keys. This author has been unable to learn why it had not been referred to as a Limited Two Step, but it is so rarely seen that the term is not especially important. (Currently it is used by one manufacturer as a special option on existing systems only.)

MASTER	<u>548210</u>
CHANGE KEYS	215543
	215545

In this format, the numbering system once again determines the "best" cut depths to get the most change keys.

NUMBERING SYSTEM	"WORST CUTS"	"BEST CUTS"
0 to 9	5	0,9
1 to 0	6	1,0
0 to 6	2,4,5	0,1,3,6
1 to 7	3,5,6	1,2,4,7
0 to 8	2,4,5,6	0,1,3,7,8
1 to 9	3,5,6,7	1,2,4,8,9
1 to 6	NOT APPLICABLE	NOT APPLICABLE

RANDOM INCREMENTS - Here, some of the positions will be in one format, and the remaining positions will be in a different format. This is considered by most master keying locksmiths to be a risky option, because it is easy to make a mistake. (Currently it is used by one manufacturer on one of its product lines due to manufacturing tolerances in that particular product to compensate for salt spray or bad weather conditions.)

MASTER	<u>548210</u>
CHANGE KEYS	335531
	335532

The same guidelines for optimum master key cuts will apply to each position as apply to the format it uses.

OF COURSE, EACH OF THESE OPTIONS WOULD HAVE FAR MORE TO ITS CHANGE KEYS THAN THE TWO SHOWN IN EACH OF THESE EXAMPLES.

With that out of the way, our next selection is the TMK (Top Master Key) bitting.

There are some general guidelines that we'd LIKE to follow on master key selection. Theoretically, any key combination could be chosen for the highest master key. But in reality, certain choices are less desirable than others, and should be avoided if at all possible.

Ideally, some good guidelines for all master keys for standard pin tumbler locks are:

1) No more than three cuts of the same depth should be used next to each other on a key (or anywhere on a 5 pin tumbler key).

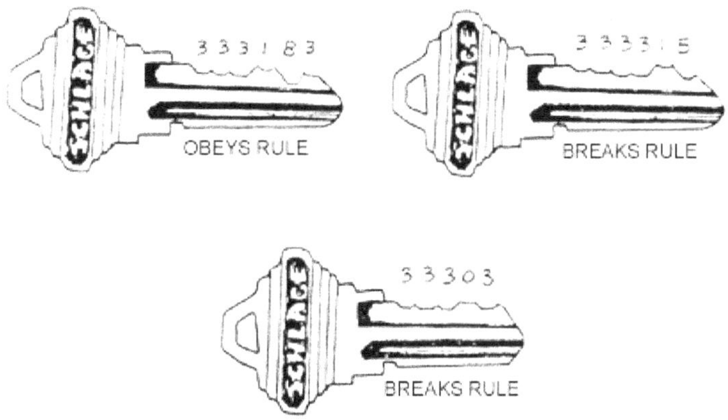

2) No more than 4 cuts of the same depth anywhere on a six pin tumbler key.

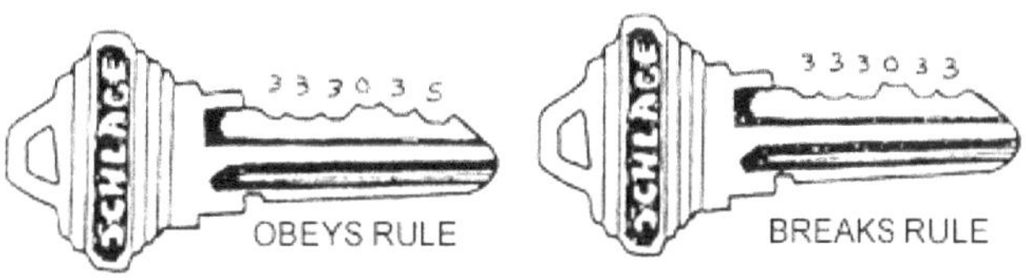

3) At least one of the cuts (though preferably not in the tip position) should be of the shallowest possible depth (most metal left on the key) so that other keys are less likely to be filed down into a master key.

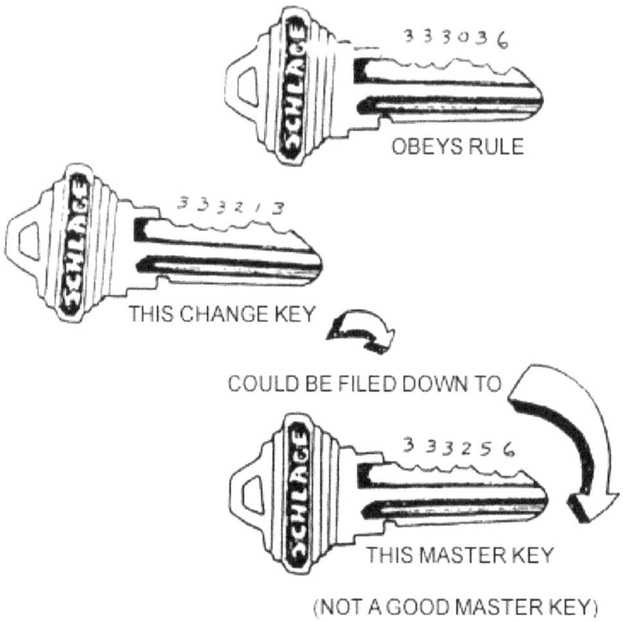

4) It is generally agreed to be a good idea to have at least one very deep cut as well, so that the bottom pins in each master keyed lock will not be too similar.

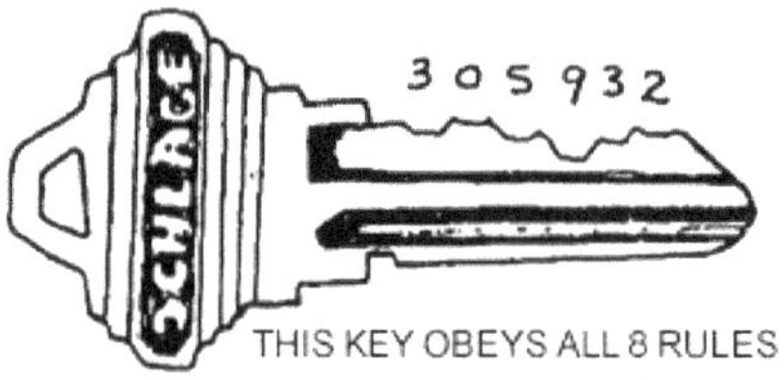

5) It should not have all high or all low cuts, because these are too easily picked.

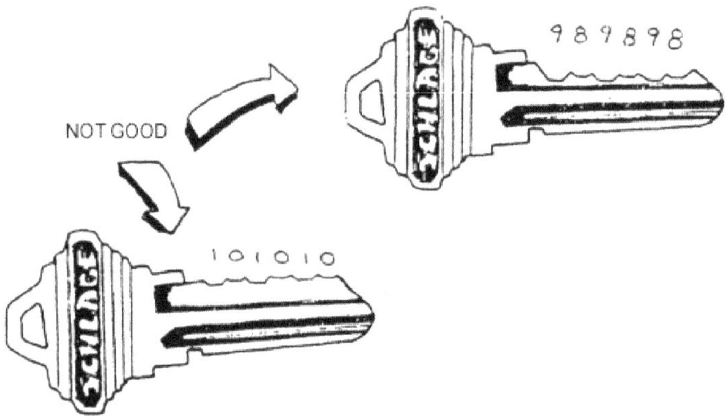

6) The cut nearest the bow of the key should not be extremely deep, since it makes the key have a tendency to break off in the lock.

7) A STAIRCASE (or DECLINING STEP) KEY should be avoided since it can frequently be pulled out of the lock in the wrong position, leaving the lock cylinder in an "unlocked" condition, where the plug can be easily turned with a screwdriver or similar tool.

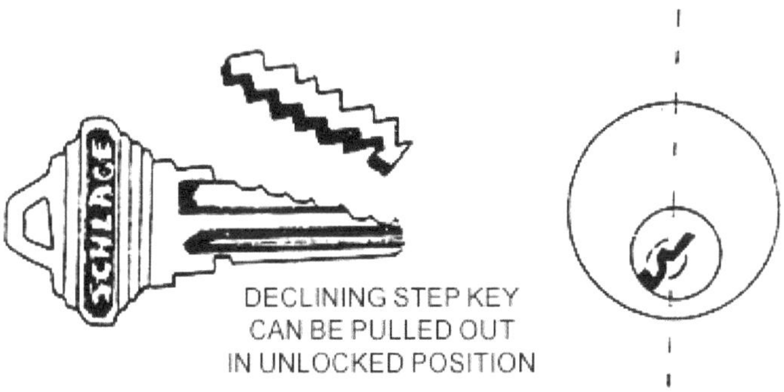

8) The cuts on the key should be within the guidelines for the MACS (Maximum Adjacent Cut Specification). This term refers to keys that we can safely assume will work when properly cut, and inserted in properly pinned lock cylinders. Older terms for this (no longer considered acceptable) are **ADJACENT BITTING TOLERANCE, SAFETY FACTOR,** and **MAXIMUM ADJACENT CUT DIFFERENTIAL**. These have been abbreviated as **ABT, SF,** and **MACD**.

The **LIST Council** liked the term **MAXIMUM ADJACENT CUT SPECIFICATION (MACS)** and the idea of pinning to the **MACS**. This author sometimes wonders if they saw the humor in their choice of terms to adopt for the industry dictionary.

Related Terms from the LIST Council Dictionary:

maximum adjacent cut specification
n. the maximum allowable difference between adjacent cut depths

In any case, the term describes a simple physical fact. Some keys cannot be properly cut and will not work if they are.

As an illustration, let's say that we are using a code machine that has a .375" thick cutting wheel, and a 90 degree full V angle. This means that the distance from the center of the cutting wheel to one edge of it would be approximately .187". Most keys on the market today have a spacing increment (center of cut to center of cut) of .156". When this cutting wheel has cut into the key blank and taken off approximately .110", the edge of the cut will be centered on the next space's centerline--right where the pin tumbler should sit. If we take off any more metal in that cut, it will begin to take off some of the metal from the cut next to it, if that cut is at a full blank height (such as the shallowest cut in some locks would be).

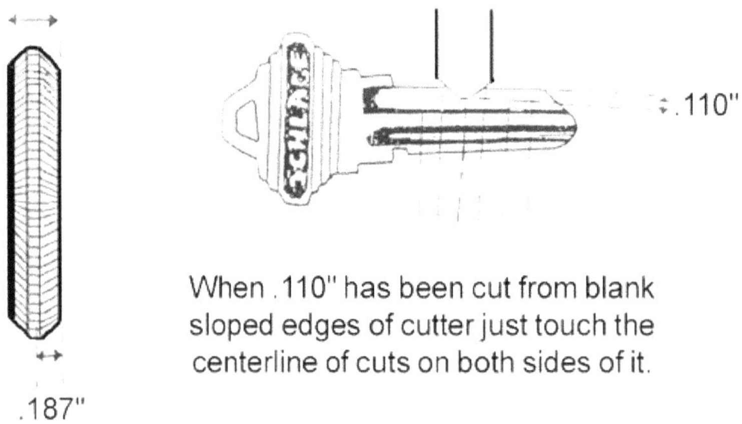

When .110" has been cut from blank sloped edges of cutter just touch the centerline of cuts on both sides of it.

To understand this more fully, let's look at a practical example of the MACS in action. Let's say that we have a Schlage Classic pin tumbler key that starts out with a full blank width of .337". A Zero cut (0) in the first space would leave .335" on the key.

This is because Schlage's Classic pin tumbler key depths are:

CUT#	METAL LEFT ON KEY
0	.335"
1	.320"
2	.305"
3	.290"
4	.275"
5	.260"
6	.245"
7	.230"
8	.215"
9	.200"

Now let's go to the second space and cut a #7 depth, which leaves .230" on the key. We have actually taken off (.337" - .230" =) a total of .107", which means that we could go .005" deeper before our cutting wheel actually TOUCHED the center of the Zero cut next to it, so the key still works. But if we had made a number eight cut in the second position, this would leave .215" on the key, meaning that we have taken off .122"---twelve thousandths of an inch deeper than where the cutting wheel just touched the centerline of the zero cut in the first position. Thus, instead of being supported by .335" of metal left on the key, the pin tumbler is only supported by .323", and the key will not work properly, if at all!!! a number Nine cut in the second position would take off .137", which is twenty-seven thousandths of an inch deeper than the acceptable .110", and now the zero cut (which should be .335") is only .308", almost two cut depths deeper than it should be.

Thus, we could say that, on a Schlage pin tumbler key, a zero cut next to a seven cut is fine, but a zero cut next to an eight or nine cut is not! A different way of expressing this relationship is to say that, in adjacent cuts (cuts next to each other on the key) a seven increment difference (7-0=7) is okay, but that any number of increments difference that is more then seven apart in adjacent cuts is NOT okay. This is referred to as a MACS of 7.

Using this safety factor of seven for Schlage, we can see that an eight cut next to a one cut is okay (8-1=7), but a nine cut next to a one cut is NOT okay (9-1=8, which is more than our safety factor).

There can be no 100% accurate chart of the MACS, because there are so many possible cutting wheel thicknesses, angles, and key blank heights, each of which will affect this.

The following chart is a rough guideline, however, that should serve most purposes.

SOME COMMON MACS ARE:

ARROW (STANDARD	**6**
ALL SFIC (SMALL FORMAT INTERCHANGEABLE CORES)	**9**
CORBIN (ALL).	**7**
DEXTER	**7**
ILCO (KABA-ILCO)	**7**
KWIKSET	**6**
LOCKWOOD	**7**
RUSSWIN	**7**
SARGENT (WITH 76 DEGREE CUTTER)	**8**
SARGENT (WITH 90 DEGREE CUTTER)	**7**
SARGENT (WITH 100 DEGREE CUTTER)	**5**
SCHLAGE (except SFIC)	**7**
WEISER	**7**
YALE .019	**7**

As we saw, the MACS will affect our possible keys for the highest master key, also, being the eighth of the "universal rules" for master key selection.

Now, quickly review the eight rules. Have we ruled out ALL the possible master keys yet? NO?? You mean there ARE a few left?

Please remember that these are only guidelines, not absolute rules, with the exception of the deep cut next to the bow of the key, the Staircase (declining step) Key, and the MACS These three are absolute, because they can cause immediate, severe, and very real problems. The others are all advisable guidelines for reasons of security, reduced wear, etc. Of course, they should be followed whenever possible.

Once we have selected our master key bittings, they should be written in the appropriate section of our chart, marked **TOP (THEORETICAL) MASTER,** and the cut matching each from the possible depths section should be crossed off, as shown here:

3	0	5	9	3	2
0	0	0	0	0	0
1	1	1	1	1	1
2	2	2	2	2	2
3	3	3	3	3	3
4	4	4	4	4	4
5	5	5	5	5	5
6	6	6	6	6	6
7	7	7	7	7	7
8	8	8	8	8	8
9	9	9	9	9	9
0	0	0	0	0	0

Because our numbering system is 0 to 9, as we saw earlier, unless there is an unusual wear pattern or poor tolerances within the lock cylinder (or core) for some reason, it is likely that this numbering system will use a two increment drop (also called a two step progression). That means if our master key cut is even, we will use no odd numbers. If our master key cut is odd we will use no even numbers. (Zero is considered an even number in master keying.)

> **Related Terms from the LIST Council Dictionary:**
>
> **two-step progression**
> n. a progression using a two increment difference between bittings of a given position

Therefore, our next step is to cross off all the cuts we will not be using for each position, as shown here:

3	0	5	9	3	2
0	0	0	0	0	0
1	1	1	1	1	1
2	2	2	2	2	2
3	3	3	3	3	3
4	4	4	4	4	4
5	5	5	5	5	5
6	6	6	6	6	6
7	7	7	7	7	7
8	8	8	8	8	8
9	9	9	9	9	9
0	0	0	0	0	0

Note that if we remove the parts that were crossed out, what we have left looks like:

3	0	5	9	3	2	TOP (THEORETICAL) MASTER
1	2	1	1	1	0	
5	4	3	3	5	4	(KBA)
7	6	7	5	7	6	KEY BITTING
9	8	9	7	9	8	ARRAY

⇦ SEQUENCE OF PROGRESSION

If you notice how the numbers in the possible depths section line up, each position goes from shallowest to deepest. That is referred to as **TOP DOWN KBA ORDER**, and was the standard until 1982 when ALOA sponsored the **National Task Group for Certified Training Programs, Master Keying Study Group** (which later became the **LIST Council**) and asked it to develop a master keying course to be called MK101.

Related Terms from the LIST Council Dictionary:

top-down order
n. an arrangement of potential key cut depths within a KBA beginning at the shallowest and continuing to the deepest

The group, which included Gerry Finch, Kenny Ehrenreich, A.J. Hoffman, Billy B. Edwards, Jerome Andrews and others, determined that the **TOP DOWN KBA ORDER** resulted in pages of undesirable keys at the front and back of the complete chart of bittings, and was not as useful as it ideally should have been.

They developed the **ODOMETER KBA ORDER**, which saved some of the steps required by the Wiegend Jensen/Bert Michaels method of filling out the **KBA**.

Step one: Write the TMK bitting in the appropriate portion of the KBA, as shown here:

3	0	5	9	3	2

TOP (THEORETICAL) MASTER

(KBA)
KEY BITTING
ARRAY

⇦ SEQUENCE OF PROGRESSION

Step Two: In each position, beginning with the number shown on the master key for that position, write the next number greater than the master key, with the same parity (Odd or Even). When you reach the highest such number rotate to the shallowest matching parity number (like the odometer on the car does), as shown here:

3	0	5	9	3	2
5	2	7	1	5	4
7	4	9	3	7	6
9	6	1	5	9	8
1	8	3	7	1	0

TOP (THEORETICAL) MASTER

(KBA)
KEY BITTING
ARRAY

⇦ SEQUENCE OF PROGRESSION

Notice how quickly this technique allows you to fill out the KBA with very little difficulty?

Related Terms from the LIST Council Dictionary:

odometer order:
n. A key bitting array order in which each position of the KBA lists the potential cut depths beginning at the next cut deeper than the master key cut, and when it reaches the deepest cut, if there are additional possible depths to be listed, begins again at the shallowest and continues from there.

It took the name **ODOMETER ORDER** from its similarity to an automotive odometer. Note how each of the possibilities ends up resembling these illustrations of the odometer wheels:

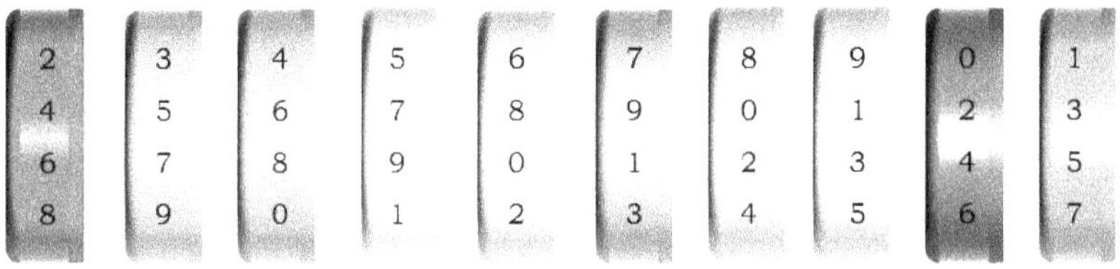

Illustration by Max Spiwak, reprinted with permission of artist.

And, similar to the odometer wheel, when we reach the highest value given, the next becomes the lowest matching value. So, looking at the second wheel from the left, If we were rolling it from a 7, it would go to the 9, then to the 1 and finally to the 3.

Still, understanding where the numbers came from originally can help make it seem even easier, which is why this book first showed you the technique developed by Wiegend Jensen and Bert Michaels.

What if our numbering system were not 0 to 9 or 1 to 0?

Lets look at a 1 to 6 using the ODOMETER KBA ORDER.

Earlier we saw this usually uses a single increment drop (or Single Step Progression).

Related Terms from the LIST Council Dictionary:

single step progression
n. a progression using a one increment difference between bittings of a given position

The first thing we see is that there are not enough rows in the standard KBA for us to put all of the numbers except the TMK numbers. (There is no parity in a standard single increment system, so it becomes ALL the cuts except the TMK cuts.)

1	3	5	6	3	2

TOP (THEORETICAL) MASTER

(KBA)
KEY BITTING
ARRAY

⇦ SEQUENCE OF PROGRESSION

Therefore, this chart will be slightly different (or we could just decide not to use one of the cuts for each space and use THIS chart) as shown here:

1	3	5	6	3	2

TOP (THEORETICAL) MASTER

(KBA)
KEY BITTING
ARRAY

⇦ SEQUENCE OF PROGRESSION

This allows us to fill in the chart using the **ODOMETER KBA ORDER**, as shown on the next page.

1	3	5	6	3	2
2	4	6	1	4	3
3	5	1	2	5	4
4	6	2	3	6	5
5	1	3	4	1	6
6	2	4	5	2	1

TOP (THEORETICAL) MASTER

(KBA)
KEY BITTING
ARRAY

⇦ SEQUENCE OF PROGRESSION

Finally we have to decide which of the System Development Rules we will apply to this particular system.

Because this book will be covering all three of the system development rules normally associated with standard pin tumbler cylinders, we will begin with the **TPP** or **Total Position Progression.**

Lesson Five: Total Position Progression

Progression is nothing more than a method of counting. If you can count, you can do Progression. For our purposes, we will be breaking up the Progression into columns called **PROGRESSION SEQUENCES**, each of which will represent a cut position on the key (and later a chamber in the lock).

Let's look at one page of a finished master keying chart. This chart design is called the **STANDARD PROGRESSION FORMAT** chart or **SPF**.

3	2	1	0	5	8	THEORETICAL MASTER
5	4	3	2	7	0	BASE FIELD (KBA) KEY BITTING ARRAY
7	6	5	4	9	2	
9	8	7	6	1	4	
1	0	9	8	3	6	
f	e	d	a	b	c	⇦ SEQUENCE OF PROGRESSION

5	4	7	2	7	0
5	4	7	4	7	0
5	4	7	6	7	0
5	4	7	8	7	0

5	4	7	2	7	2
5	4	7	4	7	2
5	4	7	6	7	2
5	4	7	8	7	2

5	4	7	2	7	4
5	4	7	4	7	4
5	4	7	6	7	4
5	4	7	8	7	4

5	4	7	2	7	6
5	4	7	4	7	6
5	4	7	6	7	6
5	4	7	8	7	6

5	4	7	2	9	0
5	4	7	4	9	0
5	4	7	6	9	0
5	4	7	8	9	0

5	4	7	2	9	2
5	4	7	4	9	2
5	4	7	6	9	2
5	4	7	8	9	2

5	4	7	2	9	4
5	4	7	4	9	4
5	4	7	6	9	4
5	4	7	8	9	4

5	4	7	2	9	6
5	4	7	4	9	6
5	4	7	6	9	6
5	4	7	8	9	6

5	4	7	2	1	0
5	4	7	4	1	0
5	4	7	6	1	0
5	4	7	8	1	0

5	4	7	2	1	2
5	4	7	4	1	2
5	4	7	6	1	2
5	4	7	8	1	2

5	4	7	2	1	4
5	4	7	4	1	4
5	4	7	6	1	4
5	4	7	8	1	4

5	4	7	2	1	6
5	4	7	4	1	6
5	4	7	6	1	6
5	4	7	8	1	6

5	4	7	2	3	0
5	4	7	4	3	0
5	4	7	6	3	0
5	4	7	8	3	0

5	4	7	2	3	2
5	4	7	4	3	2
5	4	7	6	3	2
5	4	7	8	3	2

5	4	7	2	3	4
5	4	7	4	3	4
5	4	7	6	3	4
5	4	7	8	3	4

5	4	7	2	3	6
5	4	7	4	3	6
5	4	7	6	3	6
5	4	7	8	3	6

If you are at all familiar with the **Standard Progression Format (SPF)** chart such as this, you will probably notice that we have erased some things on the chart, specifically the various master key levels. This was done for clarity in the learning stages.

Before you read any farther, take about a minute and see if you can tell what happened. Each Block has four six pin bittings. Which of the six positions in each is the First Progression Sequence (the first position on the key to have changes occur)? (Hint:

ALWAYS Read DOWN the bittings in each column for every position.) Which is the Second? And the Third? How many numbers are there per each Block?

What did we do when we ran out of numbers in the Second Progression sequence? Why are the change key bittings grouped into four VERTICAL COLUMNS?

Did you figure it out? Let's see. The position third from the right is our **FIRST PROGRESSION SEQUENCE**, because it changed from each number to the next. There are four bittings per Block, and in that position each is different. There were four numbers in the KBA in that position, (2, 4, 6 and 8) which is why they were able to create four bittings in each block. If there had been five numbers in the KBA in that position (such as we usually find in Single Increment Progression) there would have been five bittings per block.

The second position from the right in our example above is our Second Progression sequence, because it was the next to change. We used each number there with the four numbers from a Block in the First Progression sequence. Each number in that position stayed the same within the block but changed moving down to the next block beneath it.

Because there were only 4 numbers in the **SECOND PROGRESSION SEQUENCE** (7, 9, 1 and 3), once we had used each of those with a Block of four bittings from the First Progression sequence, we ran out of new numbers. This was, in this example, after 4 Blocks, because we had 4 numbers to use with one Block each. When we run out of digits in a position, we move to the next grouping and repeat them there.

We wanted some way to mark the fact that we had begun repeating, so we divided them up into VERTICAL COLUMNS of four Blocks each per column. The reason there are FOUR blocks per column is that we had four numbers in our KBA in that position. If there had been five numbers in the KBA in that position (such as we usually find in Single Increment Progression) there would have been five blocks in each column instead of four.

Because there were four possibilities in the **THIRD PROGRESSION SEQUENCE** (0, 2, 4, and 6 in the far right position in this example) there are also four columns because we are using each of those numbers with the column of numbers from our second progression sequence. This full grouping is referred to as a **PAGE**, and it typically represents the logical grouping of bittings that fits on a standard sheet of paper, making up one page of the SPF Chart. If there had been five numbers in the KBA in that position (such as we usually find in Single Increment Progression) there would have been five columns of bittings on each page instead of four.

It is important to understand that when we first create a system we have "editorial control". We decide which position gets progressed first, second and third. But once we make that selection decision, particularly in TPP or Total Position Progression, it will not change again throughout that particular system.

The rules for progression won't change, no matter WHAT position we choose to progress in what order.

The first progression sequence will always change from one number to the next.

There will always be as many numbers per block as there are different numbers to choose from in the first progression sequence.

Each number in the second progression sequence will always be used with an entire block of numbers from the first progression sequence, and there will always be as many blocks per VERTICAL COLUMN as there are different numbers to choose from in the second progression sequence.

Each number from the third progression sequence will always be used with an entire Vertical Column from the second progression sequence. There will be as many Vertical Columns as there are different numbers to choose from in the third progression sequence.

Some positions on the key may not have been progressed in our chart, by the way. When a column hasn't had any changes occur in it yet, we say that it has been "**HELD CONSTANT".** In our example above, if this were our entire system, the method would not be Total Position Progression (because our developmental rule for that is that every position on the key is progressed and differs from the TMK - Top Master Key – cuts) and the remaining three possible progression sequences would have been held constant.

However, if you compare the KBA in the upper left with the pattern of cuts in the remaining three positions, you might notice that in the fourth position from the right (third from the left) the KBA has a 3 and a 5 BEFORE it gets to the 7 which is used on this chart. Therefore, this is NOT the only chart in the system. It is at least the third chart, because the 3 and the 5 needed to be used - for a full **PAGE** each (as the 7 is used here) – before the progression would be ready for the 7 we see here.

To this point, we have seen some of the fundamental rules of progression. We have also seen that while numbers naturally progress from right to left, we can also manipulate them in other progression patterns if we choose to do so, and it does not make it any more difficult to do, because the rules of progression remain the same.

On our charts, we separated them into logical groups made up of BLOCKS and VERTICAL COLUMNS. As we can see in the **FOURTH PROGRESSION SEQUENCE**, we use each number from the KBA with a full page of bittings, so in his case we had a full page using a 3, a full page using a 5, then this page using a 7, and followed by a page using a 9.

If we were only progressing four positions, that would give us four pages, because our KBA had four numbers to use in that section of the chart. That grouping is referred to as a Group.

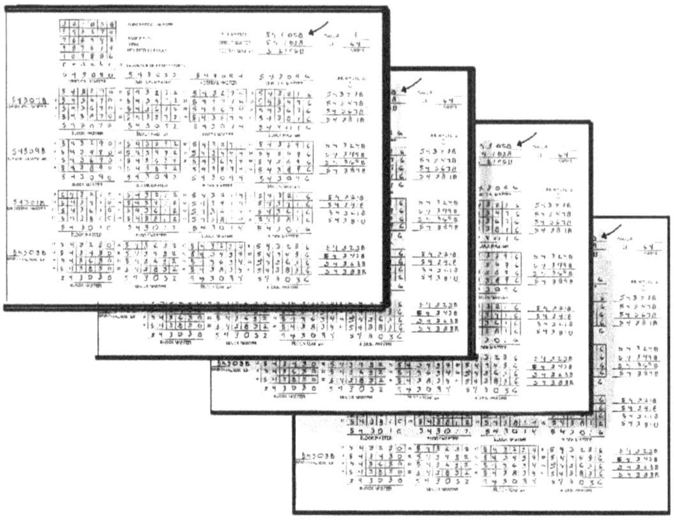

If there had been five numbers there, such as we saw might occur in Single Increment Drop, then we would have had five pages instead when we completed our progression in the fourth progression sequence. That grouping is still referred to as a Group.

Whether it is a four page or five page group depends on the number of possibilities in the fourth progression sequence (the fourth to have changes occur in the progression).

Progression sequence five will yield as many **GROUPS** as there are possibilities in that position of the KBA, and forms what is called a **SECTION** of the system.

Progression sequence six will yield as many **SECTIONS** as there are possibilities in the sixth position of the KBA, and will form a **SIX PIN SYSTEM**.

If the lock cylinders were seven pin and used seven pin keys, Progression Sequence Seven will yield as many **SIX PIN SYSTEMS** as there are possibilities in the seventh position of the KBA, and will form a **SEVEN PIN SYSTEM**.

So what decides which is progression sequence one, two, etc., in a real master key system? You, the developer of the system, do. If you are a beginner, you might find that the **ODOMETER METHOD** (right to left) is an easy to understand selection. When you get a little more experience, the second choice usually made is a Left to Right progression.

Both of these are simple because it is easy to keep track of where each progression sequence is. But manufacturers typically do not use either of those. The most common pattern used by manufacturers starts at the third from the right and works toward the right, then picks up at the fourth from the right and works toward the left, as we saw in our example above.

To keep track of which position is progressed in which order, the bottom of the KBA typically shows a place to list the Sequence of Progression. At one time it was common practice to use numbers there, but it often led to errors as those numbers were accidentally included in the layout of the SPF as though they were cut possibilities See this illustration and you will see how easy it would be to do that:

1	3	5	6	3	2	TOP (THEORETICAL) MASTER
2	4	6	1	4	3	
3	5	1	2	5	4	(KBA)
4	6	2	3	6	5	KEY BITTING ARRAY
5	1	3	4	1	6	
6	2	4	5	2	1	
1	2	3	4	5	6	⇦ SEQUENCE OF PROGRESSION

For that reason it has become common practice among a large portion of the trade for savvy master keying locksmiths to use lower case letters instead. The SKCS uses UPPER CASE LETTERS, so using them to identify the Sequence Of Progression would be confusing as well. But there is no application for lower case letters in master keying except for the Sequence Of Progression, making it an ideal choice.

Notice that in the SPF master key chart we examined, the SOP was described as:

f e d a b c

which matches what we saw when we examined the chart itself.

Let's review what we learned by the examination of that system:

The system is broken into logical groups. These logical groups are built from the Cut Possibilities Section of our Key Biting Array (KBA). If there are four possibilities in every position, four will be our multiplier, and will be reflected throughout the SPF chart.

3	0	5	9	3	2	TOP (THEORETICAL) MASTER
5	2	7	1	5	4	
7	4	9	3	7	6	(KBA)
9	6	1	5	9	8	KEY BITTING
1	8	3	7	1	0	ARRAY
						⇦ SEQUENCE OF PROGRESSION

Columns	SIXTH PROG. SEQ.	FIFTH PROG. SEQ.	FOURTH PROG. SEQ.	THIRD PROG. SEQ.	SECOND PROG. SEQ.	FIRST PROG. SEQ.
Use each digit this many times	1024	256	64	16	4	1
TBR*	4096	1024	256	64	16	4

* Total before repeating, for that column.

There will be four bittings in each block, reflecting the four possibilities in the first progression sequence.

5	4	3	2	7	0
5	4	3	4	7	0
5	4	3	6	7	0
5	4	3	8	7	0

There will be four blocks in each column, representing each of the four possibilities in the second progression sequence being used with each of the blocks developed by progression sequence one. This creates 16 bittings per column.

5	4	3	2	7	0
5	4	3	4	7	0
5	4	3	6	7	0
5	4	3	8	7	0

5	4	3	2	9	0
5	4	3	4	9	0
5	4	3	6	9	0
5	4	3	8	9	0

5	4	3	2	1	0
5	4	3	4	1	0
5	4	3	6	1	0
5	4	3	8	1	0

5	4	3	2	3	0
5	4	3	4	3	0
5	4	3	6	3	0
5	4	3	8	3	0

There will be four columns on each page, representing using each of the four possibilities in the third progression sequence with the column created by the second progression sequence. This creates 64 bittings per page.

3	2	1	0	5	8	THEORETICAL MASTER
5	4	3	2	7	0	
7	6	5	4	9	2	BASE FIELD (KBA) KEY BITTING ARRAY
9	8	7	6	1	4	
1	0	9	8	3	6	
f	e	d	a	b	c	⇦ SEQUENCE OF PROGRESSION

5	4	3	2	7	0
5	4	3	4	7	0
5	4	3	6	7	0
5	4	3	8	7	0

5	4	3	2	7	2
5	4	3	4	7	2
5	4	3	6	7	2
5	4	3	8	7	2

5	4	3	2	7	4
5	4	3	4	7	4
5	4	3	6	7	4
5	4	3	8	7	4

5	4	3	2	7	6
5	4	3	4	7	6
5	4	3	6	7	6
5	4	3	8	7	6

5	4	3	2	9	0
5	4	3	4	9	0
5	4	3	6	9	0
5	4	3	8	9	0

5	4	3	2	9	2
5	4	3	4	9	2
5	4	3	6	9	2
5	4	3	8	9	2

5	4	3	2	9	4
5	4	3	4	9	4
5	4	3	6	9	4
5	4	3	8	9	4

5	4	3	2	9	6
5	4	3	4	9	6
5	4	3	6	9	6
5	4	3	8	9	6

5	4	3	2	1	0
5	4	3	4	1	0
5	4	3	6	1	0
5	4	3	8	1	0

5	4	3	2	1	2
5	4	3	4	1	2
5	4	3	6	1	2
5	4	3	8	1	2

5	4	3	2	1	4
5	4	3	4	1	4
5	4	3	6	1	4
5	4	3	8	1	4

5	4	3	2	1	6
5	4	3	4	1	6
5	4	3	6	1	6
5	4	3	8	1	6

5	4	3	2	3	0
5	4	3	4	3	0
5	4	3	6	3	0
5	4	3	8	3	0

5	4	3	2	3	2
5	4	3	4	3	2
5	4	3	6	3	2
5	4	3	8	3	2

5	4	3	2	3	4
5	4	3	4	3	4
5	4	3	6	3	4
5	4	3	8	3	4

5	4	3	2	3	6
5	4	3	4	3	6
5	4	3	6	3	6
5	4	3	8	3	6

There will be four pages in each group, representing using each of the four possibilities in the fourth progression sequence with the page created by the third progression sequence. This creates 256 bittings per group.

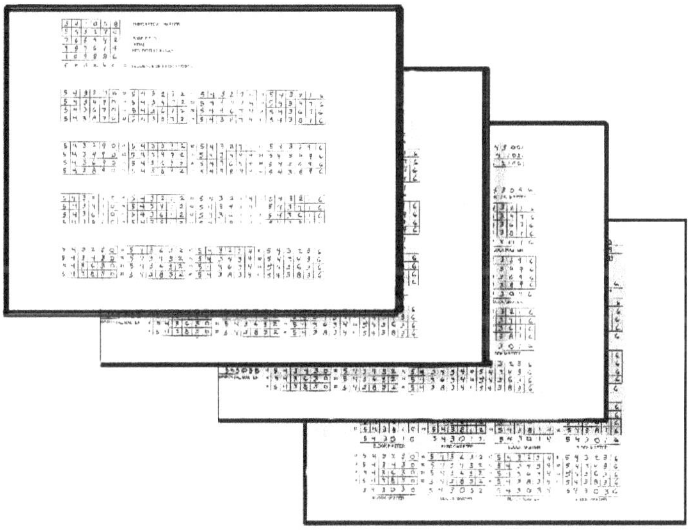

There will be four groups in each section, representing using each of the four possibilities in the fifth progression sequence with the group created by the fourth progression sequence. This creates 1,024 bittings per section.

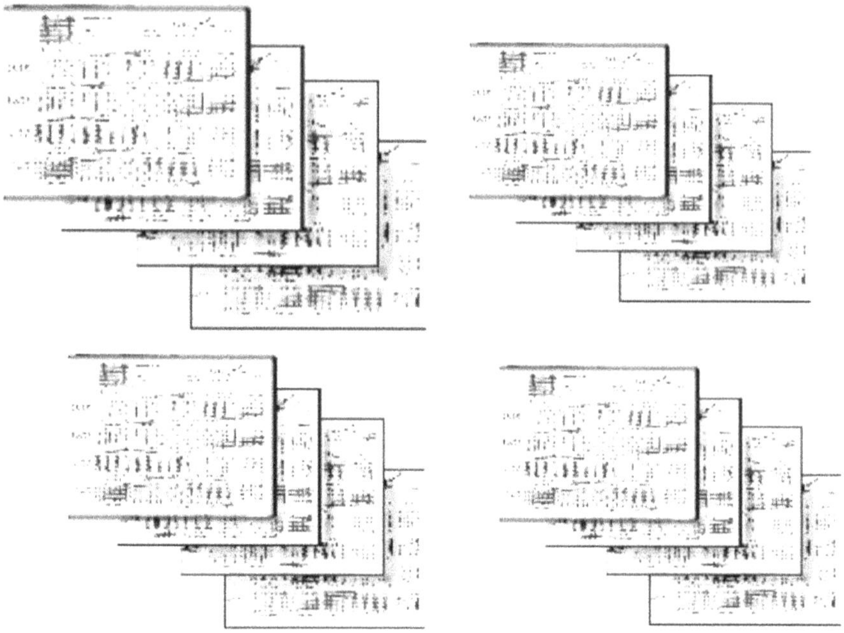

There will be four sections in each 6 pin system, representing using each of the four possibilities in the sixth progression sequence with the section created by the fifth progression sequence. This creates 4,096 bittings per 6 pin system

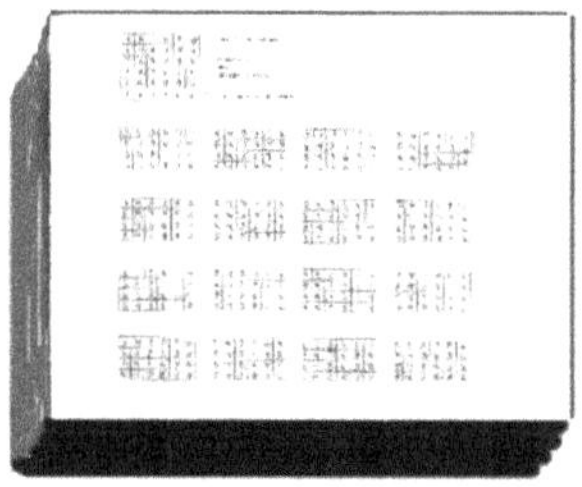

If we had seven pin cylinders, we could take it to the next level, once again using 4 as the multiplier. There will be four 6 pin systems in the 7 pin system, representing using each of the four possibilities in the sevenths progression sequence with the 6 pin system created by the sixth progression sequence. This creates 16,384 bittings per 7 pin system.

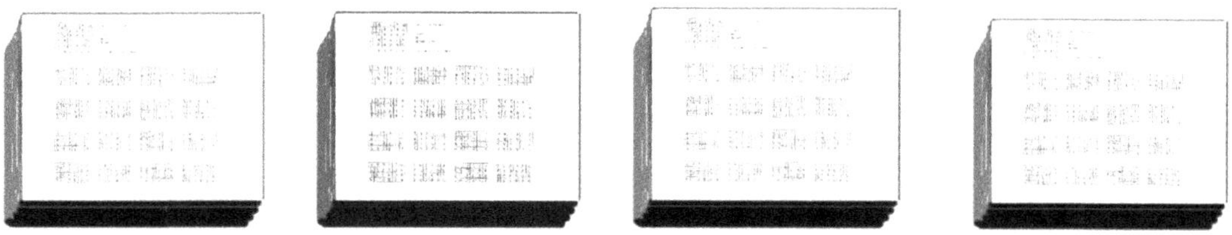

All of the above is, of course based on having a KBA with four cut possibilities in each position.

As we have mentioned here repeatedly, lock cylinders whose manufacturers have chosen to use 0 to 5 or 1 to 6 numbering systems usually have large enough increments to permit the use of a single increment of difference between keys in the system, or as it is usually called Single Increment Drop. We also have seen that this typically results in a KBA with 5 possibilities in each position. That means 5 becomes our multiplier instead of 4 throughout the system.

There will be five bittings per Block. (5 bittings per Block)
There will be five blocks per Column. (25 bittings per Column)
There will be five columns per Page. (125 bittings per Page)
There will be five Pages per Group. (625 bittings per Group)
There will be five Groups per Section (or 5 pin System). (3125 bittings)
There will be five Sections per 6 pin System (15,625 bittings)
There will be five 6 pin Systems per 7 pin System. (78,125 bittings)

Columns	SIXTH PROG. SEQ.	FIFTH PROG. SEQ.	FOURTH PROG. SEQ.	THIRD PROG. SEQ.	SECOND PROG. SEQ.	FIRST PROG. SEQ.
Use each digit this many times	3125	625	125	25	5	1
TBR*	15,625	3125	625	125	25	5

* Total before repeating, for that column.

Regardless of which chart we use, a few simple rules govern progression as we develop the master key system SPF charts:

1. In the FIRST PROGRESSION SEQUENCE the squares in each BLOCK will be identical with those from that position on the KBA Chart.

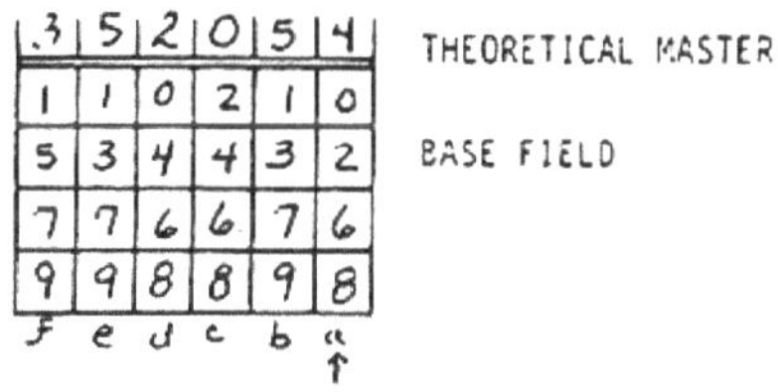

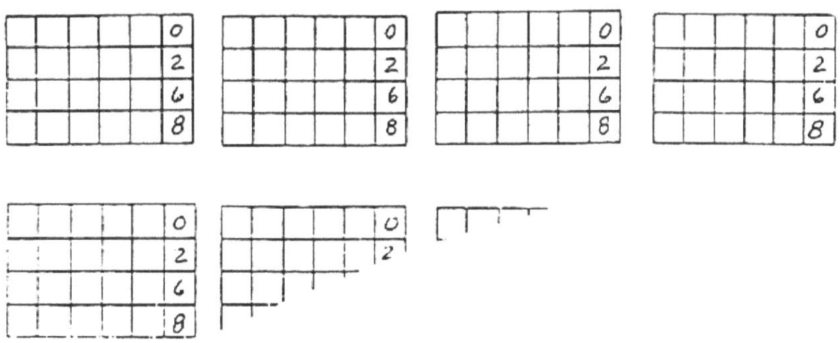

This is true regardless of options selected, as shown with a three step system in the following example:

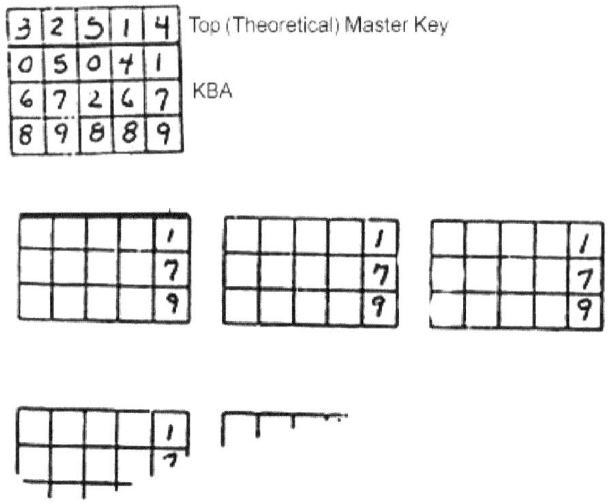

It is also true in Sequences of Progression other than simple right to left, as we saw in our earlier example and in the following illustration:

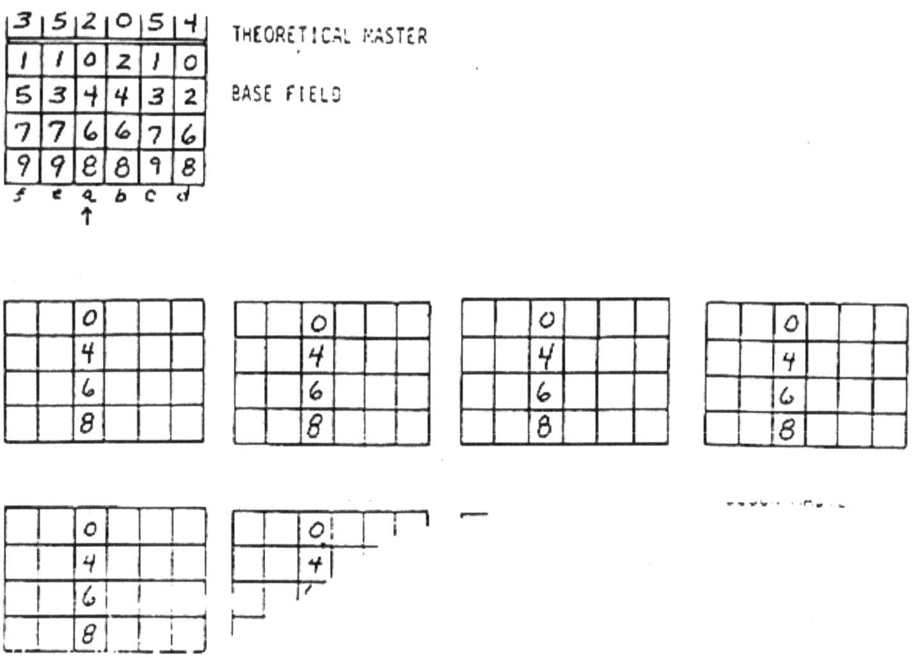

2. In the SECOND PROGRESSION SEQUENCE each BLOCK will use one of the numbers from that position of the KBA Chart, and there will be as many BLOCKS in each VERTICAL COLUMN as there are choices in that position on the KBA.

After each column, the numbers repeat in the same sequence.

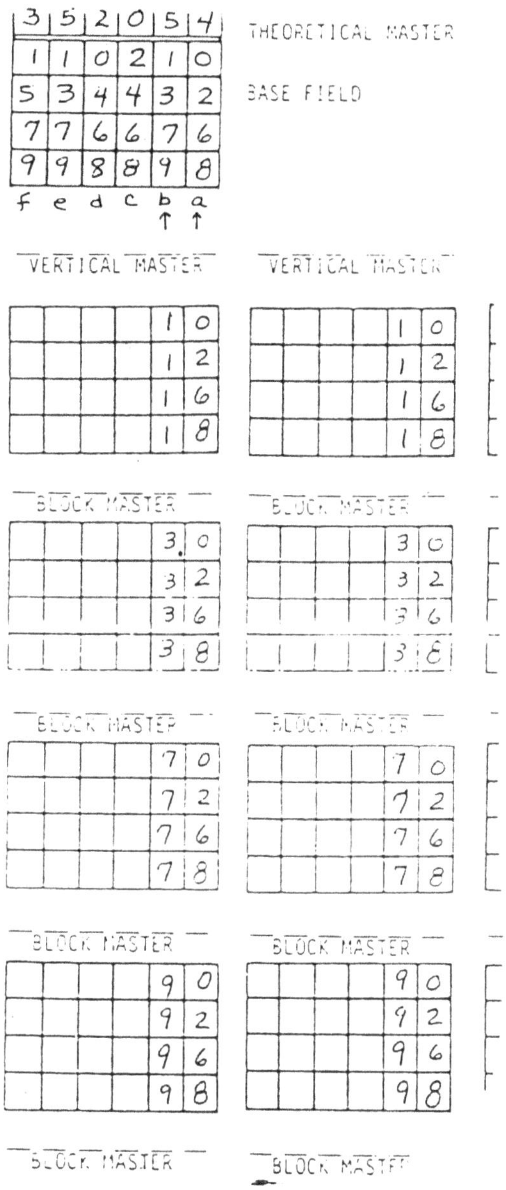

And, of course, this also holds true regardless of options selected or Sequence of Progression:

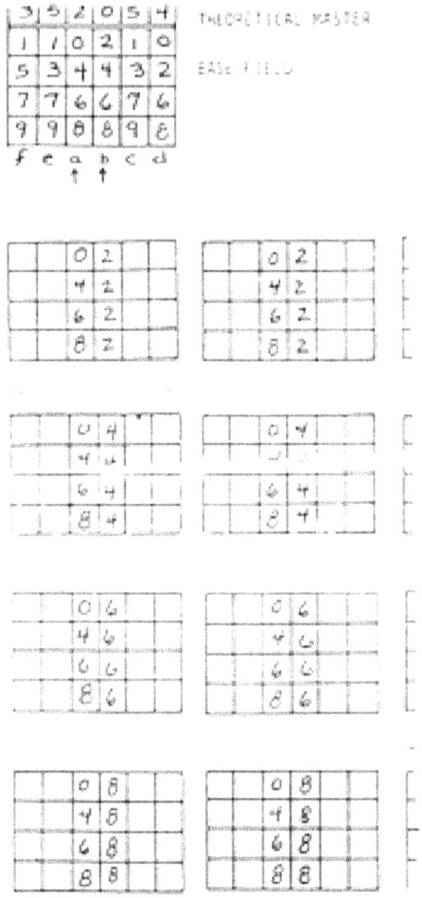

3. In the THIRD PROGRESSION SEQUENCE each of the numbers from that position on the KBA Chart will be used for all the squares in that position in a VERTICAL COLUMN, and there will be as many Vertical Columns as there are possible choices in that position of the KBA:

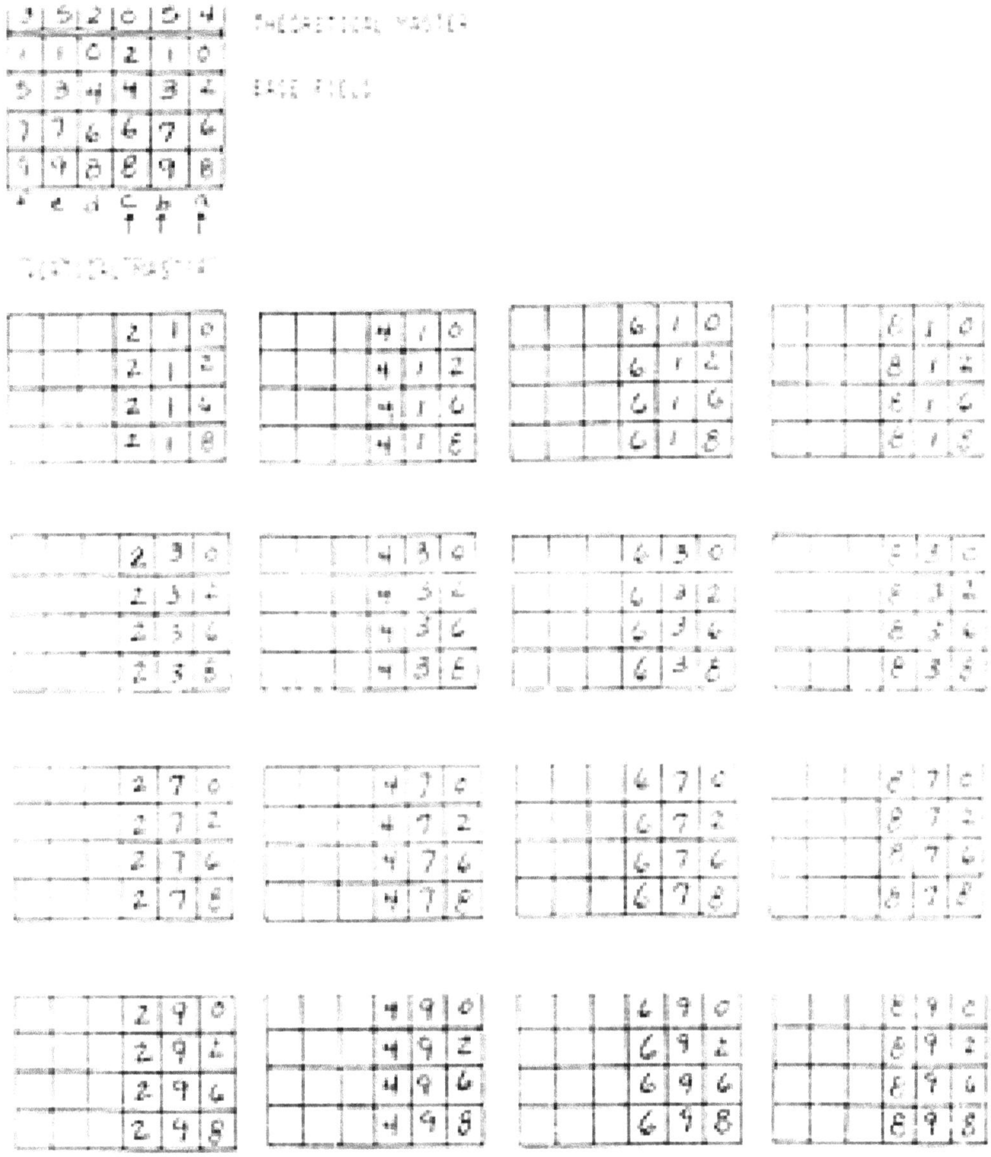

Once again, this is also true in other Sequences Of Progression:

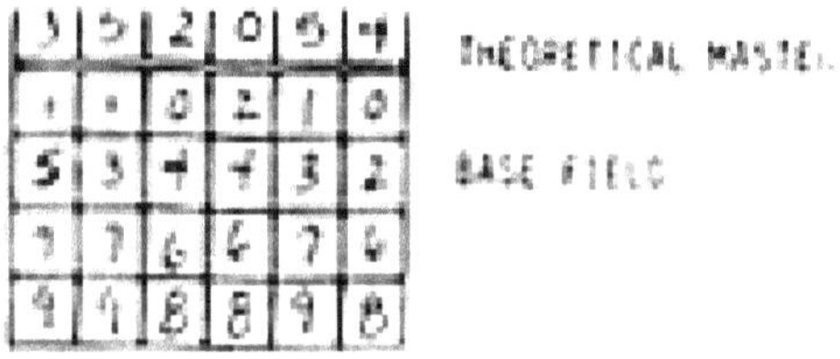

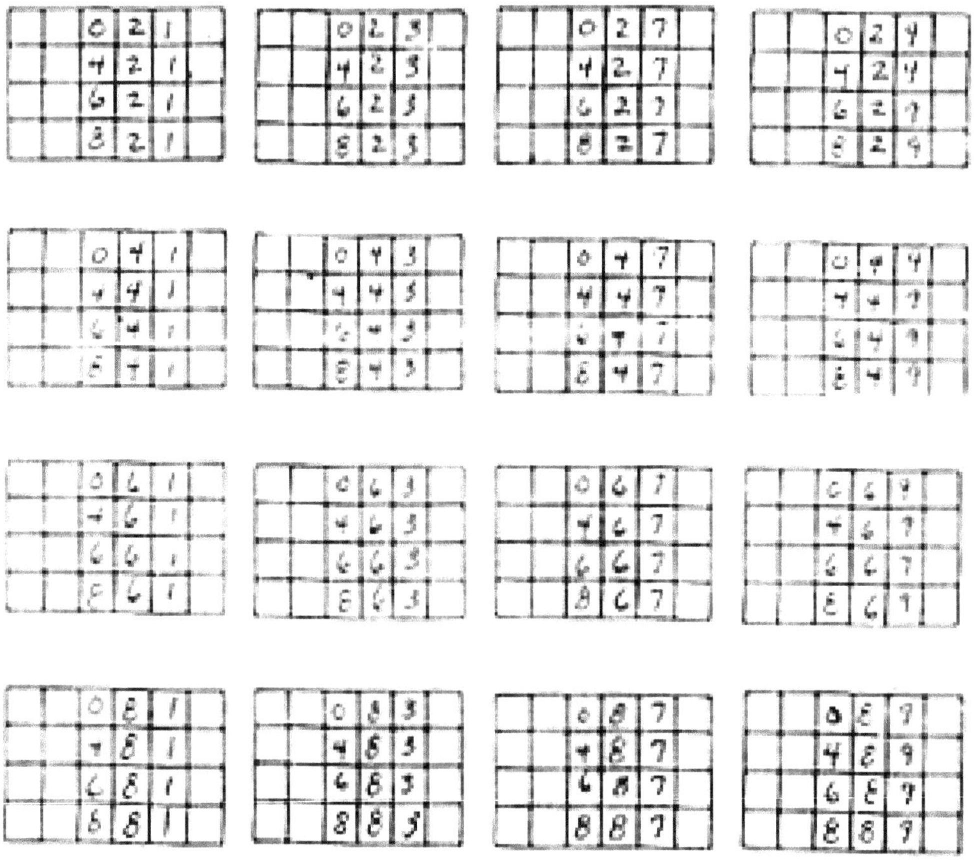

At this point in our progression, an interesting thing happens. From this point onward, all the other pages in the system will be identical in the first three sequences.

→**TIP:** Therefore, an interesting and useful idea is to take the page we have just created to a Quick Print Shop or Xerographic Process Copying Machine, and run off as many copies as your system will require to complete its progression.

YOUR NEXT GOAL IS TO NUMBER THE PAGES. THIS IS VITAL TO THE CORRECT DEVELOPMENT OF THE REMAINDER OF THE SYSTEM.

One page out of sequence during the development process from this stage on can destroy the system's usefulness.

After numbering the pages, you are ready to develop the remainder of the progression in the fourth, fifth, and sixth progression sequences.

1. **IN THE FOURTH PROGRESSION SEQUENCE, each number from that position of the KBA will be used with an entire PAGE of combinations, and there will be as many pages per GROUP as there are possible choices in that position of the KBA.**

2. **IN THE FIFTH PROGRESSION SEQUENCE, each number from that position of the KBA will be used with one GROUP of pages, and there will be as many groups of pages as there are numbers in that position of the KBA. In a 5 pin lock cylinder, this is all the progression that is possible.**

3. **IN THE SIXTH PROGRESSION SEQUENCE, each number from that position of the KBA is used with all the numbers created by the fifth progression sequence. In a six pin cylinder, this is all the progression that is possible.**

→**Tip:** If it seemed simple until you got to the 4th, 5th, and 6th progression sequences, don't despair. There is an easy method of completing the progression.

On a separate chart or piece of paper, progress the fourth, fifth, and sixth positions as though they were the first three, and use each one of the resulting combinations with an entire page of the charts, as shown on the next page.

This technique will save you much time, and greatly reduce the possibility of error.

Two rules that you MUST follow:

1. BE NEAT!!! Mistakes occur when you're not!

2. ALWAYS RUSH SLOWLY AND CAREFULLY!!! (Every job can be done faster by taking the time to do it right the first time around, avoiding a return trip.)

3	2	1	0	5	8
5	4	3	2	7	0
7	6	5	4	9	2
9	8	7	6	1	4
1	0	9	8	3	6
f	e	d			

THEORETICAL MASTER

BASE FIELD
(KBA)
KEY BITTING ARRAY

⇦ SEQUENCE OF PROGRESSION

PAGE MASTER ______
GROUP MASTER ______
SECTION MASTER ______

PAGE # ______
OF ______
PAGES

LINE MASTERS ⇩

VERTICAL MASTER | VERTICAL MASTER | VERTICAL MASTER | VERTICAL MASTER

	#							#							#							#						
	1	5	4	3				17	7	4	3				33	9	4	3				49	1	4	3			
	2	5	4	5				18	7	4	5				34	9	4	5				50	1	4	5			
HORIZONTAL MASTER	3	5	4	7				19	7	4	7				35	9	4	7				51	1	4	7			
	4	5	4	9				20	7	4	9				36	9	4	9				52	1	4	9			

BLOCK MASTER | BLOCK MASTER | BLOCK MASTER | BLOCK MASTER

	#							#							#							#						
	5	5	6	3				21	7	6	3				37	9	6	3				53	1	6	3			
	6	5	6	5				22	7	6	5				38	9	6	5				54	1	6	5			
HORIZONTAL MASTER	7	5	6	7				23	7	6	7				39	9	6	7				55	1	6	7			
	8	5	6	9				24	7	6	9				40	9	6	9				56	1	6	9			

BLOCK MASTER | BLOCK MASTER | BLOCK MASTER | BLOCK MASTER

	#							#							#							#						
	9	5	8	3				25	7	8	3				41	9	8	3				57	1	8	3			
	10	5	8	5				26	7	8	5				42	9	8	5				58	1	8	5			
HORIZONTAL MASTER	11	5	8	7				27	7	8	7				43	9	8	7				59	1	8	7			
	12	5	8	9				28	7	8	9				44	9	8	9				60	1	8	9			

BLOCK MASTER | BLOCK MASTER | BLOCK MASTER | BLOCK MASTER

	#							#							#							#						
	13	5	0	3				29	7	0	3				45	9	0	3				61	1	0	3			
	14	5	0	5				30	7	0	5				46	9	0	5				62	1	0	5			
HORIZONTAL MASTER	15	5	0	7				31	7	0	7				47	9	0	7				63	1	0	7			
	16	5	0	9				32	7	0	9				48	9	0	9				64	1	0	9			

BLOCK MASTER | BLOCK MASTER | BLOCK MASTER | BLOCK MASTER

Together, these two charts represent 64 pages, as you will see later.

3	2	1	0	5	8
5	4	3	2	7	0
7	6	5	4	9	2
9	8	7	6	1	4
1	0	9	8	3	6
f	e	d	c	b	a

THEORETICAL MASTER

BASE FIELD
(KBA)
KEY BITTING ARRAY

⇦ SEQUENCE OF PROGRESSION

PAGE MASTER ______
GROUP MASTER ______
SECTION MASTER ______

PAGE # ______
OF ______
PAGES

LINE MASTERS ⇩

VERTICAL MASTER | VERTICAL MASTER | VERTICAL MASTER | VERTICAL MASTER

	#							#							#							#						
	1				2	7	0	17				4	7	0	33				6	7	0	49				8	7	0
	2				2	7	2	18				4	7	2	34				6	7	2	50				8	7	2
HORIZONTAL MASTER	3				2	7	4	19				4	7	4	35				6	7	4	51				8	7	4
	4				2	7	6	20				4	7	6	36				6	7	6	52				8	7	6

BLOCK MASTER | BLOCK MASTER | BLOCK MASTER | BLOCK MASTER

	#							#							#							#						
	5				2	9	0	21				4	9	0	37				6	9	0	53				8	9	0
	6				2	9	2	22				4	9	2	38				6	9	2	54				8	9	2
HORIZONTAL MASTER	7				2	9	4	23				4	9	4	39				6	9	4	55				8	9	4
	8				2	9	6	24				4	9	6	40				6	9	6	56				8	9	6

BLOCK MASTER | BLOCK MASTER | BLOCK MASTER | BLOCK MASTER

	#							#							#							#						
	9				2	1	0	25				4	1	0	41				6	1	0	57				8	1	0
	10				2	1	2	26				4	1	2	42				6	1	2	58				8	1	2
HORIZONTAL MASTER	11				2	1	4	27				4	1	4	43				6	1	4	59				8	1	4
	12				2	1	6	28				4	1	6	44				6	1	6	60				8	1	6

BLOCK MASTER | BLOCK MASTER | BLOCK MASTER | BLOCK MASTER

	#							#							#							#						
	13				2	3	0	29				4	3	0	45				6	3	0	61				8	3	0
	14				2	3	2	30				4	3	2	46				6	3	2	62				8	3	2
HORIZONTAL MASTER	15				2	3	4	31				4	3	4	47				6	3	4	63				8	3	4
	16				2	3	6	32				4	3	6	48				6	3	6	64				8	3	6

BLOCK MASTER | BLOCK MASTER | BLOCK MASTER | BLOCK MASTER

Lesson Six:
OUR SAMPLE SYSTEM

Up till now, we have not worked with a real SPF chart. Our charts have been simplified to make it easier to see exactly what was happening. But if we are going to build a sample system, that no longer makes sense because at some point we will need to fill in the rest of the chart, and the simplified chart lacks the place to put the rest of what we will need.

Therefore, it is time to see a true SPF chart and to identify what was missing.

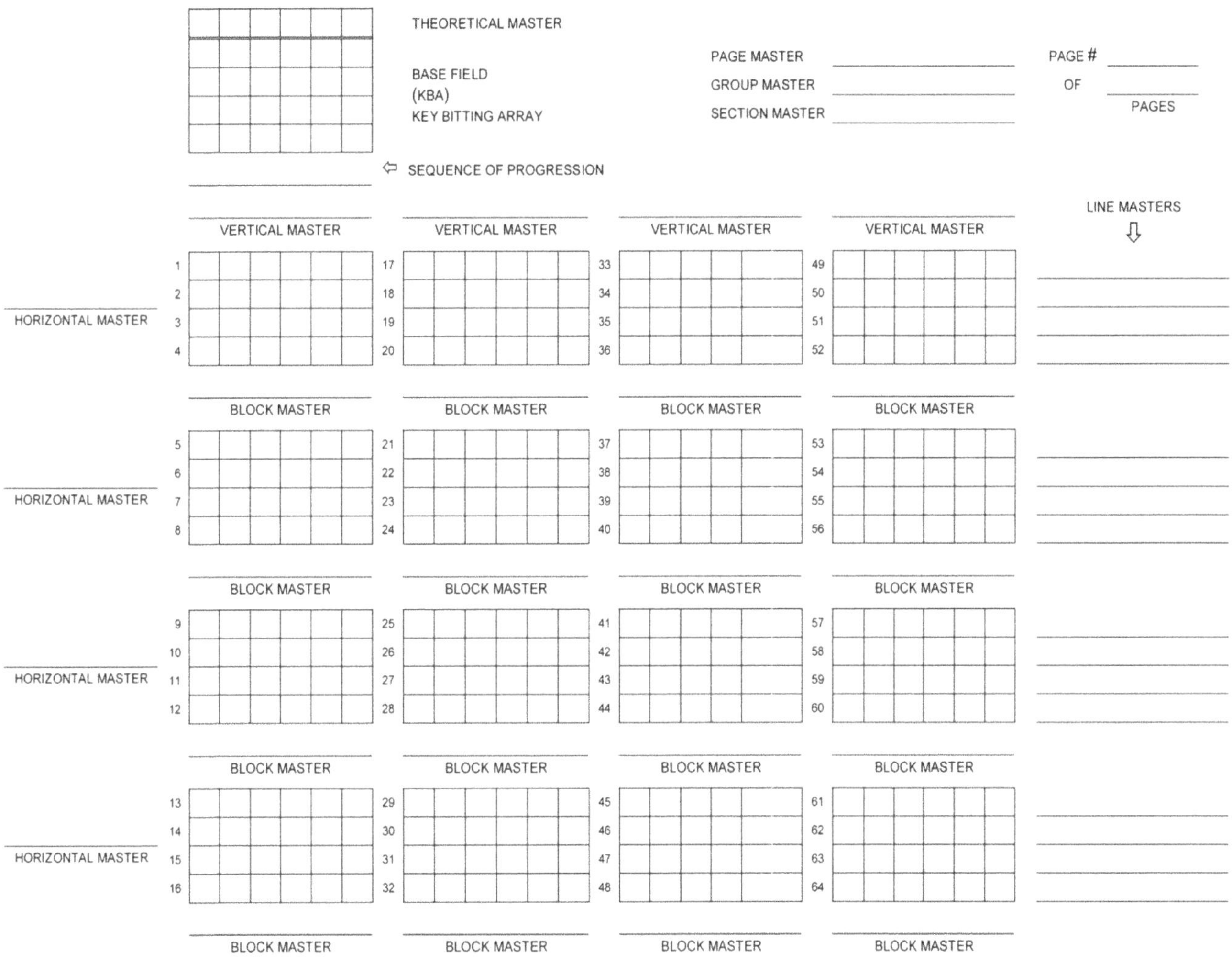

Notice that we have a few things added. In the very top right are spaces for the **Page Master, Group Master** and **Section Master** and spaces to mark the **page numbers**. Above each column is a space marked **Vertical Master** and down the left side are spaces marked **Horizontal Master**. Down the right side are a series of spaces marked at the top as **Line Masters**. Finally, under each Block is a space marked **Block Master**.

Later we will see all of those spaces put to use, but for now just remember that when we are progressing change key bittings for the system we ignore them.

Our first job is to select a numbering system and then a TMK.

We will use 3 2 1 0 5 8 with a 0 to 9 numbering system for our sample in the book, but if you are following along on a blank set of SPF charts, you can select any valid TMK. To make it as easy to see the changes occur as possible, we will use the ODOMETER METHOD for our Sequence Of Progression, or f e d c b a, so that it progresses quite logically from right to left.

For our a (FIRST Progression Sequence), bring down the numbers exactly as they appear in the KBA, one square per number, with all four numbers in the a position being used in each block. Repeat this for every block on the page.

For our b (SECOND Progression Sequence), use each of the four numbers from the b position of the KBA in each **Block**, going **DOWN** the columns. Repeat as necessary.

For our c position (THIRD Progression Sequence), use each of the four numbers from the c position of the KBA for all the blocks in a column, changing numbers with each column.

The result should look like the following charts:

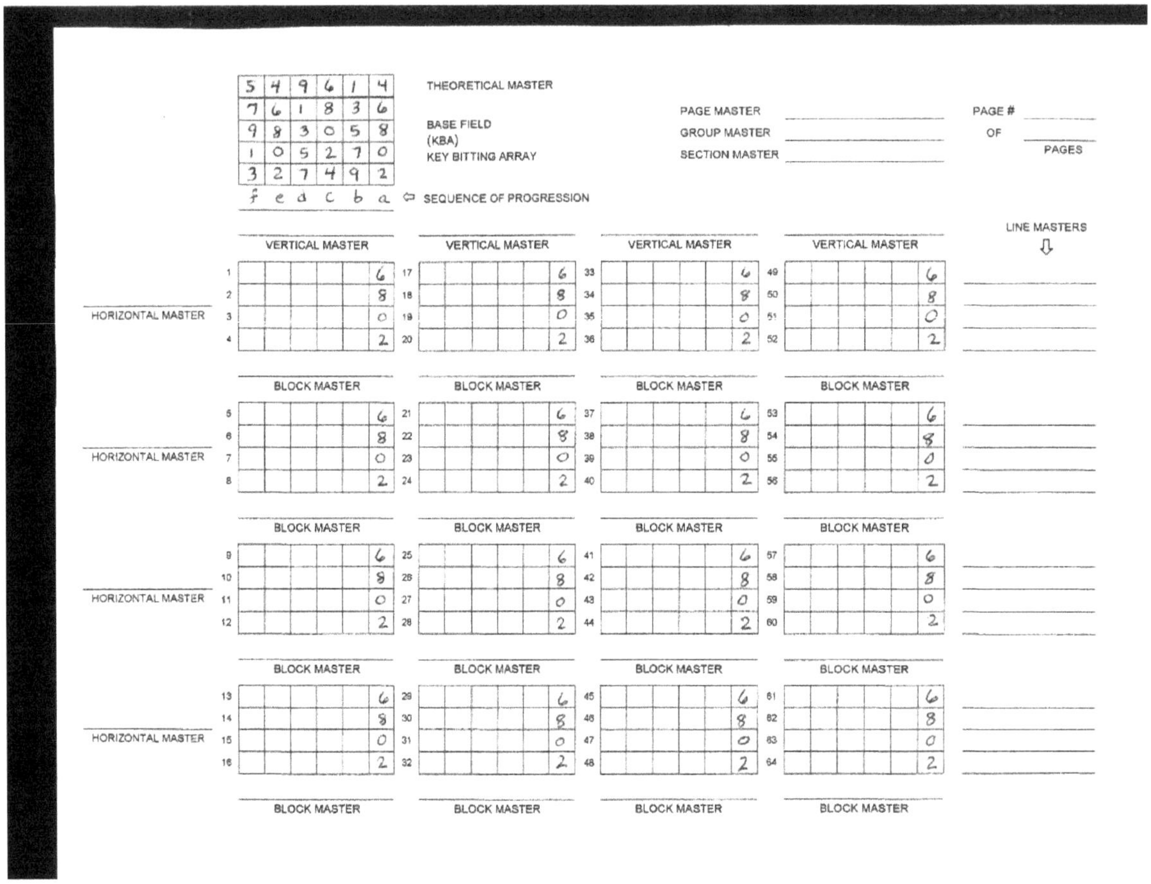
THEORETICAL MASTER

BASE FIELD (KBA) KEY BITTING ARRAY

5	4	9	6	1	4
7	6	1	8	3	6
9	8	3	0	5	8
1	0	5	2	7	0
3	2	7	4	9	2
f	e	d	c	b	a

⇦ SEQUENCE OF PROGRESSION

PAGE MASTER
GROUP MASTER
SECTION MASTER
PAGE # OF PAGES

LINE MASTERS ⇩

VERTICAL MASTER | VERTICAL MASTER | VERTICAL MASTER | VERTICAL MASTER

1	6	17	6	33	6	49	6
2	8	18	8	34	8	50	8
3	0	19	0	35	0	51	0
4	2	20	2	36	2	52	2
5	6	21	6	37	6	53	6
6	8	22	8	38	8	54	8
7	0	23	0	39	0	55	0
8	2	24	2	40	2	56	2
9	6	25	6	41	6	57	6
10	8	26	8	42	8	58	8
11	0	27	0	43	0	59	0
12	2	28	2	44	2	60	2
13	6	29	6	45	6	61	6
14	8	30	8	46	8	62	8
15	0	31	0	47	0	63	0
16	2	32	2	48	2	64	2

HORIZONTAL MASTER (beside each group of four rows)

BLOCK MASTER (below each block)

f	e	d	c	b	a
5	4	9	6	1	4
7	6	1	8	3	6
9	8	3	0	5	8
1	0	5	2	7	0
3	2	7	4	9	2
f	e	d	c	b	a

⇦ SEQUENCE OF PROGRESSION

THEORETICAL MASTER

BASE FIELD (KBA) KEY BITTING ARRAY

PAGE MASTER ______ GROUP MASTER ______ SECTION MASTER ______

PAGE # ______ OF ______ PAGES

LINE MASTERS ⇩

	#	f	e	d	c	b	a	#	f	e	d	c	b	a	#	f	e	d	c	b	a	#	f	e	d	c	b	a	LINE MASTERS
	VERTICAL MASTER							VERTICAL MASTER							VERTICAL MASTER							VERTICAL MASTER							
	1					3	6	17					3	6	33					3	6	49					3	6	
	2					3	8	18					3	8	34					3	8	50					3	8	
HORIZONTAL MASTER	3					3	0	19					3	0	35					3	0	51					3	0	
	4					3	2	20					3	2	36					3	2	52					3	2	
	BLOCK MASTER							BLOCK MASTER							BLOCK MASTER							BLOCK MASTER							
	5					5	6	21					5	6	37					5	6	53					5	6	
	6					5	8	22					5	8	38					5	8	54					5	8	
HORIZONTAL MASTER	7					5	0	23					5	0	39					5	0	55					5	0	
	8					5	2	24					5	2	40					5	2	56					5	2	
	BLOCK MASTER							BLOCK MASTER							BLOCK MASTER							BLOCK MASTER							
	9					7	6	25					7	6	41					7	6	57					7	6	
	10					7	8	26					7	8	42					7	8	58					7	8	
HORIZONTAL MASTER	11					7	0	27					7	0	43					7	0	59					7	0	
	12					7	2	28					7	2	44					7	2	60					7	2	
	BLOCK MASTER							BLOCK MASTER							BLOCK MASTER							BLOCK MASTER							
	13					9	6	29					9	6	45					9	6	61					9	6	
	14					9	8	30					9	8	46					9	8	62					9	8	
HORIZONTAL MASTER	15					9	0	31					9	0	47					9	0	63					9	0	
	16					9	2	32					9	2	48					9	2	64					9	2	
	BLOCK MASTER							BLOCK MASTER							BLOCK MASTER							BLOCK MASTER							

f	e	d	c	b	a
5	4	9	6	1	4
7	6	1	8	3	6
9	8	3	0	5	8
1	0	5	2	7	0
3	2	7	4	9	2
f	e	d	c	b	a

⇦ SEQUENCE OF PROGRESSION

THEORETICAL MASTER

BASE FIELD (KBA) KEY BITTING ARRAY

PAGE MASTER ______ GROUP MASTER ______ SECTION MASTER ______

PAGE # ______ OF ______ PAGES

LINE MASTERS ⇩

	#	f	e	d	c	b	a	#	f	e	d	c	b	a	#	f	e	d	c	b	a	#	f	e	d	c	b	a	LINE MASTERS
	VERTICAL MASTER							VERTICAL MASTER							VERTICAL MASTER							VERTICAL MASTER							
	1				8	3	6	17				0	3	6	33				2	3	6	49				4	3	6	
	2				8	3	8	18				0	3	8	34				2	3	8	50				4	3	8	
HORIZONTAL MASTER	3				8	3	0	19				0	3	0	35				2	3	0	51				4	3	0	
	4				8	3	2	20				0	3	2	36				2	3	2	52				4	3	2	
	BLOCK MASTER							BLOCK MASTER							BLOCK MASTER							BLOCK MASTER							
	5				8	5	6	21				0	5	6	37				2	5	6	53				4	5	6	
	6				8	5	8	22				0	5	8	38				2	5	8	54				4	5	8	
HORIZONTAL MASTER	7				8	5	0	23				0	5	0	39				2	5	0	55				4	5	0	
	8				8	5	2	24				0	5	2	40				2	5	2	56				4	5	2	
	BLOCK MASTER							BLOCK MASTER							BLOCK MASTER							BLOCK MASTER							
	9				8	7	6	25				0	7	6	41				2	7	6	57				4	7	6	
	10				8	7	8	26				0	7	8	42				2	7	8	58				4	7	8	
HORIZONTAL MASTER	11				8	7	0	27				0	7	0	43				2	7	0	59				4	7	0	
	12				8	7	2	28				0	7	2	44				2	7	2	60				4	7	2	
	BLOCK MASTER							BLOCK MASTER							BLOCK MASTER							BLOCK MASTER							
	13				8	9	6	29				0	9	6	45				2	9	6	61				4	9	6	
	14				8	9	8	30				0	9	8	46				2	9	8	62				4	9	8	
HORIZONTAL MASTER	15				8	9	0	31				0	9	0	47				2	9	0	63				4	9	0	
	16				8	9	2	32				0	9	2	48				2	9	2	64				4	9	2	
	BLOCK MASTER							BLOCK MASTER							BLOCK MASTER							BLOCK MASTER							

The next logical step is to make copies of the page you have created so far and number them from 1 to 64

→**TIP:** You might want to make about 70 copies so that if you make an error you do not end up running out of charts prematurely.

→**TIP:** This is also the time to instead decide to use a two part SPF as a shortcut. If you were doing that you would skip the following steps and apply the technique instead, as shown on the next page.

(If you don't "get it" right now, don't worry. It will be much clearer soon.)

If you are not using the two chart SPF shortcut, you need to continue the progression.

Next, we progress our d position, by using each number there with a full page of charts, and repeating the pattern every 4 pages. Your D position would look like the following four charts:

THEORETICAL MASTER

BASE FIELD
(KBA)
KEY BITTING ARRAY

3	2	1	0	5	8
5	4	3	2	7	0
7	6	5	4	9	2
9	8	7	6	1	4
1	0	9	8	3	6
f	e	d	a	b	c

⇦ SEQUENCE OF PROGRESSION

PAGE MASTER ______

GROUP MASTER ______

SECTION MASTER ______

VERTICAL MASTER

1			3	2	7	0
2			3	4	7	0
3			3	6	7	0
4			3	8	7	0

BLOCK MASTER

5			3	2	9	0
6			3	4	9	0
7			3	6	9	0
8			3	8	9	0

BLOCK MASTER

9			3	2	1	0
10			3	4	1	0
11			3	6	1	0
12			3	8	1	0

BLOCK MASTER

13			3	2	3	0
14			3	4	3	0
15			3	6	3	0
16			3	8	3	0

BLOCK MASTER

VERTICAL MASTER

17			3	2	7	2
18			3	4	7	2
19			3	6	7	2
20			3	8	7	2

BLOCK MASTER

21			3	2	9	2
22			3	4	9	2
23			3	6	9	2
24			3	8	9	2

BLOCK MASTER

25			3	2	1	2
26			3	4	1	2
27			3	6	1	2
28			3	8	1	2

BLOCK MASTER

29			3	2	3	2
30			3	4	3	2
31			3	6	3	2
32			3	8	3	2

BLOCK MASTER

VERTICAL MASTER

33			3	2	7	4
34			3	4	7	4
35			3	6	7	4
36			3	8	7	4

BLOCK MASTER

37			3	2	9	4
38			3	4	9	4
39			3	6	9	4
40			3	8	9	4

BLOCK MASTER

41			3	2	1	4
42			3	4	1	4
43			3	6	1	4
44			3	8	1	4

BLOCK MASTER

45			3	2	3	4
46			3	4	3	4
47			3	6	3	4
48			3	8	3	4

BLOCK MASTER

VERTICAL MASTER

49			3	2	7	6
50			3	4	7	6
51			3	6	7	6
52			3	8	7	6

BLOCK MASTER

53			3	2	9	6
54			3	4	9	6
55			3	6	9	6
56			3	8	9	6

BLOCK MASTER

57			3	2	1	6
58			3	4	1	6
59			3	6	1	6
60			3	8	1	6

BLOCK MASTER

61			3	2	3	6
62			3	4	3	6
63			3	6	3	6
64			3	8	3	6

BLOCK MASTER

Pages 1, 5, 9, 13, 17, 21, 25, 29, 33, 37, 41, 45, 49, 53, 57, 61 look like the chart above.

3	2	1	0	5	8
5	4	3	2	7	0
7	6	5	4	9	2
9	8	7	6	1	4
1	0	9	8	3	6

f e d a b c ⇦ SEQUENCE OF PROGRESSION

THEORETICAL MASTER

BASE FIELD
(KBA)
KEY BITTING ARRAY

PAGE MASTER ____________

GROUP MASTER ____________

SECTION MASTER ____________

VERTICAL MASTER | VERTICAL MASTER | VERTICAL MASTER | VERTICAL MASTER

1			5	2	7	0	17			5	2	7	2	33			5	2	7	4	49			5	2	7	6
2			5	4	7	0	18			5	4	7	2	34			5	4	7	4	50			5	4	7	6
3			5	6	7	0	19			5	6	7	2	35			5	6	7	4	51			5	6	7	6
4			5	8	7	0	20			5	8	7	2	36			5	8	7	4	52			5	8	7	6

BLOCK MASTER | BLOCK MASTER | BLOCK MASTER | BLOCK MASTER

5			5	2	9	0	21			5	2	9	2	37			5	2	9	4	53			5	2	9	6
6			5	4	9	0	22			5	4	9	2	38			5	4	9	4	54			5	4	9	6
7			5	6	9	0	23			5	6	9	2	39			5	6	9	4	55			5	6	9	6
8			5	8	9	0	24			5	8	9	2	40			5	8	9	4	56			5	8	9	6

BLOCK MASTER | BLOCK MASTER | BLOCK MASTER | BLOCK MASTER

9			5	2	1	0	25			5	2	1	2	41			5	2	1	4	57			5	2	1	6
10			5	4	1	0	26			5	4	1	2	42			5	4	1	4	58			5	4	1	6
11			5	6	1	0	27			5	6	1	2	43			5	6	1	4	59			5	6	1	6
12			5	8	1	0	28			5	8	1	2	44			5	8	1	4	60			5	8	1	6

BLOCK MASTER | BLOCK MASTER | BLOCK MASTER | BLOCK MASTER

13			5	2	3	0	29			5	2	3	2	45			5	2	3	4	61			5	2	3	6
14			5	4	3	0	30			5	4	3	2	46			5	4	3	4	62			5	4	3	6
15			5	6	3	0	31			5	6	3	2	47			5	6	3	4	63			5	6	3	6
16			5	8	3	0	32			5	8	3	2	48			5	8	3	4	64			5	8	3	6

BLOCK MASTER | BLOCK MASTER | BLOCK MASTER | BLOCK MASTER

Pages 2, 6, 10, 14, 18, 22, 26, 30, 34, 38, 42, 46, 50, 54, 58, 62 look like the chart above.

3	2	1	0	5	8
5	4	3	2	7	0
7	6	5	4	9	2
9	8	7	6	1	4
1	0	9	8	3	6
f	e	d	a	b	c

THEORETICAL MASTER

BASE FIELD
(KBA)
KEY BITTING ARRAY

⇦ SEQUENCE OF PROGRESSION

PAGE MASTER ____________
GROUP MASTER ____________
SECTION MASTER ____________

VERTICAL MASTER | VERTICAL MASTER | VERTICAL MASTER | VERTICAL MASTER

1			7	2	7	0	17			7	2	7	2	33			7	2	7	4	49			7	2	7	6
2			7	4	7	0	18			7	4	7	2	34			7	4	7	4	50			7	4	7	6
3			7	6	7	0	19			7	6	7	2	35			7	6	7	4	51			7	6	7	6
4			7	8	7	0	20			7	8	7	2	36			7	8	7	4	52			7	8	7	6

BLOCK MASTER | BLOCK MASTER | BLOCK MASTER | BLOCK MASTER

5			7	2	9	0	21			7	2	9	2	37			7	2	9	4	53			7	2	9	6
6			7	4	9	0	22			7	4	9	2	38			7	4	9	4	54			7	4	9	6
7			7	6	9	0	23			7	6	9	2	39			7	6	9	4	55			7	6	9	6
8			7	8	9	0	24			7	8	9	2	40			7	8	9	4	56			7	8	9	6

BLOCK MASTER | BLOCK MASTER | BLOCK MASTER | BLOCK MASTER

9			7	2	1	0	25			7	2	1	2	41			7	2	1	4	57			7	2	1	6
10			7	4	1	0	26			7	4	1	2	42			7	4	1	4	58			7	4	1	6
11			7	6	1	0	27			7	6	1	2	43			7	6	1	4	59			7	6	1	6
12			7	8	1	0	28			7	8	1	2	44			7	8	1	4	60			7	8	1	6

BLOCK MASTER | BLOCK MASTER | BLOCK MASTER | BLOCK MASTER

13			7	2	3	0	29			7	2	3	2	45			7	2	3	4	61			7	2	3	6
14			7	4	3	0	30			7	4	3	2	46			7	4	3	4	62			7	4	3	6
15			7	6	3	0	31			7	6	3	2	47			7	6	3	4	63			7	6	3	6
16			7	8	3	0	32			7	8	3	2	48			7	8	3	4	64			7	8	3	6

BLOCK MASTER | BLOCK MASTER | BLOCK MASTER | BLOCK MASTER

Pages 3, 7, 11, 15, 19, 23, 27, 31, 35, 39, 43, 47, 51, 55, 59, 63 look like the chart above.

3	2	1	0	5	8
5	4	3	2	7	0
7	6	5	4	9	2
9	8	7	6	1	4
1	0	9	8	3	6
f	e	d	a	b	c

THEORETICAL MASTER

BASE FIELD
(KBA)
KEY BITTING ARRAY

⇦ SEQUENCE OF PROGRESSION

PAGE MASTER ____________
GROUP MASTER ____________
SECTION MASTER ____________

VERTICAL MASTER | VERTICAL MASTER | VERTICAL MASTER | VERTICAL MASTER

#							#							#							#						
1			9	2	7	0	17			9	2	7	2	33			9	2	7	4	49			9	2	7	6
2			9	4	7	0	18			9	4	7	2	34			9	4	7	4	50			9	4	7	6
3			9	6	7	0	19			9	6	7	2	35			9	6	7	4	51			9	6	7	6
4			9	8	7	0	20			9	8	7	2	36			9	8	7	4	52			9	8	7	6

BLOCK MASTER | BLOCK MASTER | BLOCK MASTER | BLOCK MASTER

#							#							#							#						
5			9	2	9	0	21			9	2	9	2	37			9	2	9	4	53			9	2	9	6
6			9	4	9	0	22			9	4	9	2	38			9	4	9	4	54			9	4	9	6
7			9	6	9	0	23			9	6	9	2	39			9	6	9	4	55			9	6	9	6
8			9	8	9	0	24			9	8	9	2	40			9	8	9	4	56			9	8	9	6

BLOCK MASTER | BLOCK MASTER | BLOCK MASTER | BLOCK MASTER

#							#							#							#						
9			9	2	1	0	25			9	2	1	2	41			9	2	1	4	57			9	2	1	6
10			9	4	1	0	26			9	4	1	2	42			9	4	1	4	58			9	4	1	6
11			9	6	1	0	27			9	6	1	2	43			9	6	1	4	59			9	6	1	6
12			9	8	1	0	28			9	8	1	2	44			9	8	1	4	60			9	8	1	6

BLOCK MASTER | BLOCK MASTER | BLOCK MASTER | BLOCK MASTER

#							#							#							#						
13			9	2	3	0	29			9	2	3	2	45			9	2	3	4	61			9	2	3	6
14			9	4	3	0	30			9	4	3	2	46			9	4	3	4	62			9	4	3	6
15			9	6	3	0	31			9	6	3	2	47			9	6	3	4	63			9	6	3	6
16			9	8	3	0	32			9	8	3	2	48			9	8	3	4	64			9	8	3	6

BLOCK MASTER | BLOCK MASTER | BLOCK MASTER | BLOCK MASTER

Pages 4, 8, 12, 16, 20, 24, 28, 32, 36, 40, 44, 48, 52, 56, 60, 64 look like the chart above.

In the fifth position, we will use each number from the KBA (4, 6, 8, 0) with four pages of the chart, and this will repeat the pattern after page 16.

Pages 1-4, 17-20, 33- 36 and 49-52 will use the 4 from the above KBA in the e position.
Pages 5-8, 21-24, 37-40 and 53-56 will use the 6 from the above KBA in the e position.
Pages 9-12, 25-28, 41-44 and 57-60 will use the 8 from the above KBA in the e position
Pages 13-16, 29-32, 45-48 and 61-64 will use the 0 from the above KBA in the e position

In the sixth position, we will use each number from the KBA (5, 7, 9, 1) with sixteen pages of the chart, and this will repeat the pattern after page 64 if we have a seven pin (which in our example we do not).

Pages 1-16 will use the 4 from the above KBA in the e position.
Pages 17-33 will use the 6 from the above KBA in the e position.
Pages 34-48 will use the 8 from the above KBA in the e position
Pages 49-64 will use the 0 from the above KBA in the e position

The following 17 charts will allow you to see the majority of changes that occur:

THEORETICAL MASTER

BASE FIELD (KBA) KEY BITTING ARRAY

3	2	1	0	5	8
5	4	3	2	7	0
7	6	5	4	9	2
9	8	7	6	1	4
1	0	9	8	3	6
f	e	d	c	b	a

= SEQUENCE OF PROGRESSION

PAGE MASTER ____
GROUP MASTER ____
SECTION MASTER ____

PAGE # 1 OF 64 PAGES

VERTICAL MASTER — LINE MASTERS

1	5	4	3	2	7	0	17	5	4	3	4	7	0	33	5	4	3	6	7	0	49	5	4	3	8	7	0
2	5	4	3	2	7	2	18	5	4	3	4	7	2	34	5	4	3	6	7	2	50	5	4	3	8	7	2
3	5	4	3	2	7	4	19	5	4	3	4	7	4	35	5	4	3	6	7	4	51	5	4	3	8	7	4
4	5	4	3	2	7	6	20	5	4	3	4	7	6	36	5	4	3	6	7	6	52	5	4	3	8	7	6
5	5	4	3	2	9	0	21	5	4	3	4	9	0	37	5	4	3	6	9	0	53	5	4	3	8	9	0
6	5	4	3	2	9	2	22	5	4	3	4	9	2	38	5	4	3	6	9	2	54	5	4	3	8	9	2
7	5	4	3	2	9	4	23	5	4	3	4	9	4	39	5	4	3	6	9	4	55	5	4	3	8	9	4
8	5	4	3	2	9	6	24	5	4	3	4	9	6	40	5	4	3	6	9	6	56	5	4	3	8	9	6
9	5	4	3	2	1	0	25	5	4	3	4	1	0	41	5	4	3	6	1	0	57	5	4	3	8	1	0
10	5	4	3	2	1	2	26	5	4	3	4	1	2	42	5	4	3	6	1	2	58	5	4	3	8	1	2
11	5	4	3	2	1	4	27	5	4	3	4	1	4	43	5	4	3	6	1	4	59	5	4	3	8	1	4
12	5	4	3	2	1	6	28	5	4	3	4	1	6	44	5	4	3	6	1	6	60	5	4	3	8	1	6
13	5	4	3	2	3	0	29	5	4	3	4	3	0	45	5	4	3	6	3	0	61	5	4	3	8	3	0
14	5	4	3	2	3	2	30	5	4	3	4	3	2	46	5	4	3	6	3	2	62	5	4	3	8	3	2
15	5	4	3	2	3	4	31	5	4	3	4	3	4	47	5	4	3	6	3	4	63	5	4	3	8	3	4
16	5	4	3	2	3	6	32	5	4	3	4	3	6	48	5	4	3	6	3	6	64	5	4	3	8	3	6

HORIZONTAL MASTER (one for each group of four rows); BLOCK MASTER (below each block of four)

3	2	1	0	5	8
5	4	3	2	7	0
7	6	5	4	9	2
9	8	7	6	1	4
1	0	9	8	3	6
f	e	d	c	b	a

THEORETICAL MASTER

BASE FIELD (KBA) KEY BITTING ARRAY

⇐ SEQUENCE OF PROGRESSION

PAGE MASTER ______
GROUP MASTER ______
SECTION MASTER ______

PAGE # 2 OF 64 PAGES

LINE MASTERS ⇩

	#	Bitting	#	Bitting	#	Bitting	#	Bitting
		VERTICAL MASTER		VERTICAL MASTER		VERTICAL MASTER		VERTICAL MASTER
	1	5 4 5 2 7 0	17	5 4 5 4 7 0	33	5 4 5 6 7 0	49	5 4 5 8 7 0
	2	5 4 5 2 7 2	18	5 4 5 4 7 2	34	5 4 5 6 7 2	50	5 4 5 8 7 2
HORIZONTAL MASTER	3	5 4 5 2 7 4	19	5 4 5 4 7 4	35	5 4 5 6 7 4	51	5 4 5 8 7 4
	4	5 4 5 2 7 6	20	5 4 5 4 7 6	36	5 4 5 6 7 6	52	5 4 5 8 7 6
		BLOCK MASTER		BLOCK MASTER		BLOCK MASTER		BLOCK MASTER
	5	5 4 5 2 9 0	21	5 4 5 4 9 0	37	5 4 5 6 9 0	53	5 4 5 8 9 0
	6	5 4 5 2 9 2	22	5 4 5 4 9 2	38	5 4 5 6 9 2	54	5 4 5 8 9 2
HORIZONTAL MASTER	7	5 4 5 2 9 4	23	5 4 5 4 9 4	39	5 4 5 6 9 4	55	5 4 5 8 9 4
	8	5 4 5 2 9 6	24	5 4 5 4 9 6	40	5 4 5 6 9 6	56	5 4 5 8 9 6
		BLOCK MASTER		BLOCK MASTER		BLOCK MASTER		BLOCK MASTER
	9	5 4 5 2 1 0	25	5 4 5 4 1 0	41	5 4 5 6 1 0	57	5 4 5 8 1 0
	10	5 4 5 2 1 2	26	5 4 5 4 1 2	42	5 4 5 6 1 2	58	5 4 5 8 1 2
HORIZONTAL MASTER	11	5 4 5 2 1 4	27	5 4 5 4 1 4	43	5 4 5 6 1 4	59	5 4 5 8 1 4
	12	5 4 5 2 1 6	28	5 4 5 4 1 6	44	5 4 5 6 1 6	60	5 4 5 8 1 6
		BLOCK MASTER		BLOCK MASTER		BLOCK MASTER		BLOCK MASTER
	13	5 4 5 2 3 0	29	5 4 5 4 3 0	45	5 4 5 6 3 0	61	5 4 5 8 3 0
	14	5 4 5 2 3 2	30	5 4 5 4 3 2	46	5 4 5 6 3 2	62	5 4 5 8 3 2
HORIZONTAL MASTER	15	5 4 5 2 3 4	31	5 4 5 4 3 4	47	5 4 5 6 3 4	63	5 4 5 8 3 4
	16	5 4 5 2 3 6	32	5 4 5 4 3 6	48	5 4 5 6 3 6	64	5 4 5 8 3 6
		BLOCK MASTER		BLOCK MASTER		BLOCK MASTER		BLOCK MASTER

3	2	1	0	5	8
5	4	3	2	7	0
7	6	5	4	9	2
9	8	7	6	1	4
1	0	9	8	3	6
f	e	d	c	b	a

THEORETICAL MASTER

BASE FIELD (KBA) KEY BITTING ARRAY

⇐ SEQUENCE OF PROGRESSION

PAGE MASTER ______
GROUP MASTER ______
SECTION MASTER ______

PAGE # 3 OF 64 PAGES

LINE MASTERS ⇩

	#	Bitting	#	Bitting	#	Bitting	#	Bitting
		VERTICAL MASTER		VERTICAL MASTER		VERTICAL MASTER		VERTICAL MASTER
	1	5 4 7 2 7 0	17	5 4 7 4 7 0	33	5 4 7 6 7 0	49	5 4 7 8 7 0
	2	5 4 7 2 7 2	18	5 4 7 4 7 2	34	5 4 7 6 7 2	50	5 4 7 8 7 2
HORIZONTAL MASTER	3	5 4 7 2 7 4	19	5 4 7 4 7 4	35	5 4 7 6 7 4	51	5 4 7 8 7 4
	4	5 4 7 2 7 6	20	5 4 7 4 7 6	36	5 4 7 6 7 6	52	5 4 7 8 7 6
		BLOCK MASTER		BLOCK MASTER		BLOCK MASTER		BLOCK MASTER
	5	5 4 7 2 9 0	21	5 4 7 4 9 0	37	5 4 7 6 9 0	53	5 4 7 8 9 0
	6	5 4 7 2 9 2	22	5 4 7 4 9 2	38	5 4 7 6 9 2	54	5 4 7 8 9 2
HORIZONTAL MASTER	7	5 4 7 2 9 4	23	5 4 7 4 9 4	39	5 4 7 6 9 4	55	5 4 7 8 9 4
	8	5 4 7 2 9 6	24	5 4 7 4 9 6	40	5 4 7 6 9 6	56	5 4 7 8 9 6
		BLOCK MASTER		BLOCK MASTER		BLOCK MASTER		BLOCK MASTER
	9	5 4 7 2 1 0	25	5 4 7 4 1 0	41	5 4 7 6 1 0	57	5 4 7 8 1 0
	10	5 4 7 2 1 2	26	5 4 7 4 1 2	42	5 4 7 6 1 2	58	5 4 7 8 1 2
HORIZONTAL MASTER	11	5 4 7 2 1 4	27	5 4 7 4 1 4	43	5 4 7 6 1 4	59	5 4 7 8 1 4
	12	5 4 7 2 1 6	28	5 4 7 4 1 6	44	5 4 7 6 1 6	60	5 4 7 8 1 6
		BLOCK MASTER		BLOCK MASTER		BLOCK MASTER		BLOCK MASTER
	13	5 4 7 2 3 0	29	5 4 7 4 3 0	45	5 4 7 6 3 0	61	5 4 7 8 3 0
	14	5 4 7 2 3 2	30	5 4 7 4 3 2	46	5 4 7 6 3 2	62	5 4 7 8 3 2
HORIZONTAL MASTER	15	5 4 7 2 3 4	31	5 4 7 4 3 4	47	5 4 7 6 3 4	63	5 4 7 8 3 4
	16	5 4 7 2 3 6	32	5 4 7 4 3 6	48	5 4 7 6 3 6	64	5 4 7 8 3 6
		BLOCK MASTER		BLOCK MASTER		BLOCK MASTER		BLOCK MASTER

3	2	1	0	5	8
5	4	3	2	7	0
7	6	5	4	9	2
9	8	7	6	1	4
1	0	9	8	3	6
f	e	d	c	b	a

THEORETICAL MASTER

BASE FIELD (KBA) KEY BITTING ARRAY

⇐ SEQUENCE OF PROGRESSION

PAGE MASTER ____________

GROUP MASTER ____________

SECTION MASTER ____________

PAGE # 4 OF 64 PAGES

LINE MASTERS ⇩

VERTICAL MASTER | VERTICAL MASTER | VERTICAL MASTER | VERTICAL MASTER

HORIZONTAL MASTER

#							#							#							#						
1	5	4	9	2	7	0	17	5	4	9	4	7	0	33	5	4	9	6	7	0	49	5	4	9	8	7	0
2	5	4	9	2	7	2	18	5	4	9	4	7	2	34	5	4	9	6	7	2	50	5	4	9	8	7	2
3	5	4	9	2	7	4	19	5	4	9	4	7	4	35	5	4	9	6	7	4	51	5	4	9	8	7	4
4	5	4	9	2	7	6	20	5	4	9	4	7	6	36	5	4	9	6	7	6	52	5	4	9	8	7	6

BLOCK MASTER | BLOCK MASTER | BLOCK MASTER | BLOCK MASTER

HORIZONTAL MASTER

#							#							#							#						
5	5	4	9	2	9	0	21	5	4	9	4	9	0	37	5	4	9	6	9	0	53	5	4	9	8	9	0
6	5	4	9	2	9	2	22	5	4	9	4	9	2	38	5	4	9	6	9	2	54	5	4	9	8	9	2
7	5	4	9	2	9	4	23	5	4	9	4	9	4	39	5	4	9	6	9	4	55	5	4	9	8	9	4
8	5	4	9	2	9	6	24	5	4	9	4	9	6	40	5	4	9	6	9	6	56	5	4	9	8	9	6

BLOCK MASTER | BLOCK MASTER | BLOCK MASTER | BLOCK MASTER

HORIZONTAL MASTER

#							#							#							#						
9	5	4	9	2	1	0	25	5	4	9	4	1	0	41	5	4	9	6	1	0	57	5	4	9	8	1	0
10	5	4	9	2	1	2	26	5	4	9	4	1	2	42	5	4	9	6	1	2	58	5	4	9	8	1	2
11	5	4	9	2	1	4	27	5	4	9	4	1	4	43	5	4	9	6	1	4	59	5	4	9	8	1	4
12	5	4	9	2	1	6	28	5	4	9	4	1	6	44	5	4	9	6	1	6	60	5	4	9	8	1	6

BLOCK MASTER | BLOCK MASTER | BLOCK MASTER | BLOCK MASTER

HORIZONTAL MASTER

#							#							#							#						
13	5	4	9	2	3	0	29	5	4	9	4	3	0	45	5	4	9	6	3	0	61	5	4	9	8	3	0
14	5	4	9	2	3	2	30	5	4	9	4	3	2	46	5	4	9	6	3	2	62	5	4	9	8	3	2
15	5	4	9	2	3	4	31	5	4	9	4	3	4	47	5	4	9	6	3	4	63	5	4	9	8	3	4
16	5	4	9	2	3	6	32	5	4	9	4	3	6	48	5	4	9	6	3	6	64	5	4	9	8	3	6

BLOCK MASTER | BLOCK MASTER | BLOCK MASTER | BLOCK MASTER

3	2	1	0	5	8
5	4	3	2	7	0
7	6	5	4	9	2
9	8	7	6	1	4
1	0	9	8	3	6
f	e	d	c	b	a

THEORETICAL MASTER

BASE FIELD (KBA) KEY BITTING ARRAY

⇐ SEQUENCE OF PROGRESSION

PAGE MASTER ____________

GROUP MASTER ____________

SECTION MASTER ____________

PAGE # 5 OF 64 PAGES

LINE MASTERS ⇩

VERTICAL MASTER | VERTICAL MASTER | VERTICAL MASTER | VERTICAL MASTER

HORIZONTAL MASTER

#							#							#							#						
1	5	6	3	2	7	0	17	5	6	3	4	7	0	33	5	6	3	6	7	0	49	5	6	3	8	7	0
2	5	6	3	2	7	2	18	5	6	3	4	7	2	34	5	6	3	6	7	2	50	5	6	3	8	7	2
3	5	6	3	2	7	4	19	5	6	3	4	7	4	35	5	6	3	6	7	4	51	5	6	3	8	7	4
4	5	6	3	2	7	6	20	5	6	3	4	7	6	36	5	6	3	6	7	6	52	5	6	3	8	7	6

BLOCK MASTER | BLOCK MASTER | BLOCK MASTER | BLOCK MASTER

HORIZONTAL MASTER

#							#							#							#						
5	5	6	3	2	9	0	21	5	6	3	4	9	0	37	5	6	3	6	9	0	53	5	6	3	8	9	0
6	5	6	3	2	9	2	22	5	6	3	4	9	2	38	5	6	3	6	9	2	54	5	6	3	8	9	2
7	5	6	3	2	9	4	23	5	6	3	4	9	4	39	5	6	3	6	9	4	55	5	6	3	8	9	4
8	5	6	3	2	9	6	24	5	6	3	4	9	6	40	5	6	3	6	9	6	56	5	6	3	8	9	6

BLOCK MASTER | BLOCK MASTER | BLOCK MASTER | BLOCK MASTER

HORIZONTAL MASTER

#							#							#							#						
9	5	6	3	2	1	0	25	5	6	3	4	1	0	41	5	6	3	6	1	0	57	5	6	3	8	1	0
10	5	6	3	2	1	2	26	5	6	3	4	1	2	42	5	6	3	6	1	2	58	5	6	3	8	1	2
11	5	6	3	2	1	4	27	5	6	3	4	1	4	43	5	6	3	6	1	4	59	5	6	3	8	1	4
12	5	6	3	2	1	6	28	5	6	3	4	1	6	44	5	6	3	6	1	6	60	5	6	3	8	1	6

BLOCK MASTER | BLOCK MASTER | BLOCK MASTER | BLOCK MASTER

HORIZONTAL MASTER

#							#							#							#						
13	5	6	3	2	3	0	29	5	6	3	4	3	0	45	5	6	3	6	3	0	61	5	6	3	8	3	0
14	5	6	3	2	3	2	30	5	6	3	4	3	2	46	5	6	3	6	3	2	62	5	6	3	8	3	2
15	5	6	3	2	3	4	31	5	6	3	4	3	4	47	5	6	3	6	3	4	63	5	6	3	8	3	4
16	5	6	3	2	3	6	32	5	6	3	4	3	6	48	5	6	3	6	3	6	64	5	6	3	8	3	6

BLOCK MASTER | BLOCK MASTER | BLOCK MASTER | BLOCK MASTER

3	2	1	0	5	8
5	4	3	2	7	0
7	6	5	4	9	2
9	8	7	6	1	4
1	0	9	8	3	6
f	e	d	c	b	a

THEORETICAL MASTER

BASE FIELD (KBA) KEY BITTING ARRAY

⇦ SEQUENCE OF PROGRESSION

PAGE MASTER ______

GROUP MASTER ______

SECTION MASTER ______

PAGE # 6 OF 64 PAGES

LINE MASTERS ⇩

	VERTICAL MASTER		VERTICAL MASTER		VERTICAL MASTER		VERTICAL MASTER
1	5 6 5 2 7 0	17	5 6 5 4 7 0	33	5 6 5 6 7 0	49	5 6 5 8 7 0
2	5 6 5 2 7 2	18	5 6 5 4 7 2	34	5 6 5 6 7 2	50	5 6 5 8 7 2
3	5 6 5 2 7 4	19	5 6 5 4 7 4	35	5 6 5 6 7 4	51	5 6 5 8 7 4
4	5 6 5 2 7 6	20	5 6 5 4 7 6	36	5 6 5 6 7 6	52	5 6 5 8 7 6

HORIZONTAL MASTER (rows 1–4); BLOCK MASTER (each column)

5	5 6 5 2 9 0	21	5 6 5 4 9 0	37	5 6 5 6 9 0	53	5 6 5 8 9 0
6	5 6 5 2 9 2	22	5 6 5 4 9 2	38	5 6 5 6 9 2	54	5 6 5 8 9 2
7	5 6 5 2 9 4	23	5 6 5 4 9 4	39	5 6 5 6 9 4	55	5 6 5 8 9 4
8	5 6 5 2 9 6	24	5 6 5 4 9 6	40	5 6 5 6 9 6	56	5 6 5 8 9 6

HORIZONTAL MASTER (rows 5–8); BLOCK MASTER (each column)

9	5 6 5 2 1 0	25	5 6 5 4 1 0	41	5 6 5 6 1 0	57	5 6 5 8 1 0
10	5 6 5 2 1 2	26	5 6 5 4 1 2	42	5 6 5 6 1 2	58	5 6 5 8 1 2
11	5 6 5 2 1 4	27	5 6 5 4 1 4	43	5 6 5 6 1 4	59	5 6 5 8 1 4
12	5 6 5 2 1 6	28	5 6 5 4 1 6	44	5 6 5 6 1 6	60	5 6 5 8 1 6

HORIZONTAL MASTER (rows 9–12); BLOCK MASTER (each column)

13	5 6 5 2 3 0	29	5 6 5 4 3 0	45	5 6 5 6 3 0	61	5 6 5 8 3 0
14	5 6 5 2 3 2	30	5 6 5 4 3 2	46	5 6 5 6 3 2	62	5 6 5 8 3 2
15	5 6 5 2 3 4	31	5 6 5 4 3 4	47	5 6 5 6 3 4	63	5 6 5 8 3 4
16	5 6 5 2 3 6	32	5 6 5 4 3 6	48	5 6 5 6 3 6	64	5 6 5 8 3 6

HORIZONTAL MASTER (rows 13–16); BLOCK MASTER (each column)

3	2	1	0	5	8
5	4	3	2	7	0
7	6	5	4	9	2
9	8	7	6	1	4
1	0	9	8	3	6
f	e	d	c	b	a

THEORETICAL MASTER

BASE FIELD (KBA) KEY BITTING ARRAY

⇦ SEQUENCE OF PROGRESSION

PAGE MASTER ______

GROUP MASTER ______

SECTION MASTER ______

PAGE # 7 OF 64 PAGES

LINE MASTERS ⇩

	VERTICAL MASTER		VERTICAL MASTER		VERTICAL MASTER		VERTICAL MASTER
1	5 6 7 2 7 0	17	5 6 7 4 7 0	33	5 6 7 6 7 0	49	5 6 7 8 7 0
2	5 6 7 2 7 2	18	5 6 7 4 7 2	34	5 6 7 6 7 2	50	5 6 7 8 7 2
3	5 6 7 2 7 4	19	5 6 7 4 7 4	35	5 6 7 6 7 4	51	5 6 7 8 7 4
4	5 6 7 2 7 6	20	5 6 7 4 7 6	36	5 6 7 6 7 6	52	5 6 7 8 7 6

HORIZONTAL MASTER (rows 1–4); BLOCK MASTER (each column)

5	5 6 7 2 9 0	21	5 6 7 4 9 0	37	5 6 7 6 9 0	53	5 6 7 8 9 0
6	5 6 7 2 9 2	22	5 6 7 4 9 2	38	5 6 7 6 9 2	54	5 6 7 8 9 2
7	5 6 7 2 9 4	23	5 6 7 4 9 4	39	5 6 7 6 9 4	55	5 6 7 8 9 4
8	5 6 7 2 9 6	24	5 6 7 4 9 6	40	5 6 7 6 9 6	56	5 6 7 8 9 6

HORIZONTAL MASTER (rows 5–8); BLOCK MASTER (each column)

9	5 6 7 2 1 0	25	5 6 7 4 1 0	41	5 6 7 6 1 0	57	5 6 7 8 1 0
10	5 6 7 2 1 2	26	5 6 7 4 1 2	42	5 6 7 6 1 2	58	5 6 7 8 1 2
11	5 6 7 2 1 4	27	5 6 7 4 1 4	43	5 6 7 6 1 4	59	5 6 7 8 1 4
12	5 6 7 2 1 6	28	5 6 7 4 1 6	44	5 6 7 6 1 6	60	5 6 7 8 1 6

HORIZONTAL MASTER (rows 9–12); BLOCK MASTER (each column)

13	5 6 7 2 3 0	29	5 6 7 4 3 0	45	5 6 7 6 3 0	61	5 6 7 8 3 0
14	5 6 7 2 3 2	30	5 6 7 4 3 2	46	5 6 7 6 3 2	62	5 6 7 8 3 2
15	5 6 7 2 3 4	31	5 6 7 4 3 4	47	5 6 7 6 3 4	63	5 6 7 8 3 4
16	5 6 7 2 3 6	32	5 6 7 4 3 6	48	5 6 7 6 3 6	64	5 6 7 8 3 6

HORIZONTAL MASTER (rows 13–16); BLOCK MASTER (each column)

3	2	1	0	5	8
5	4	3	2	7	0
7	6	5	4	9	2
9	8	7	6	1	4
1	0	9	8	3	6
f	e	d	c	b	a

THEORETICAL MASTER

BASE FIELD (KBA) KEY BITTING ARRAY

⇐ SEQUENCE OF PROGRESSION

PAGE MASTER ________

GROUP MASTER ________

SECTION MASTER ________

PAGE # 8 OF 64 PAGES

LINE MASTERS ⇩

VERTICAL MASTER / VERTICAL MASTER / VERTICAL MASTER / VERTICAL MASTER

#	f	e	d	c	b	a	#	f	e	d	c	b	a	#	f	e	d	c	b	a	#	f	e	d	c	b	a
1	5	6	9	2	7	0	17	5	6	9	4	7	0	33	5	6	9	6	7	0	49	5	6	9	8	7	0
2	5	6	9	2	7	2	18	5	6	9	4	7	2	34	5	6	9	6	7	2	50	5	6	9	8	7	2
3	5	6	9	2	7	4	19	5	6	9	4	7	4	35	5	6	9	6	7	4	51	5	6	9	8	7	4
4	5	6	9	2	7	6	20	5	6	9	4	7	6	36	5	6	9	6	7	6	52	5	6	9	8	7	6
5	5	6	9	2	9	0	21	5	6	9	4	9	0	37	5	6	9	6	9	0	53	5	6	9	8	9	0
6	5	6	9	2	9	2	22	5	6	9	4	9	2	38	5	6	9	6	9	2	54	5	6	9	8	9	2
7	5	6	9	2	9	4	23	5	6	9	4	9	4	39	5	6	9	6	9	4	55	5	6	9	8	9	4
8	5	6	9	2	9	6	24	5	6	9	4	9	6	40	5	6	9	6	9	6	56	5	6	9	8	9	6
9	5	6	9	2	1	0	25	5	6	9	4	1	0	41	5	6	9	6	1	0	57	5	6	9	8	1	0
10	5	6	9	2	1	2	26	5	6	9	4	1	2	42	5	6	9	6	1	2	58	5	6	9	8	1	2
11	5	6	9	2	1	4	27	5	6	9	4	1	4	43	5	6	9	6	1	4	59	5	6	9	8	1	4
12	5	6	9	2	1	6	28	5	6	9	4	1	6	44	5	6	9	6	1	6	60	5	6	9	8	1	6
13	5	6	9	2	3	0	29	5	6	9	4	3	0	45	5	6	9	6	3	0	61	5	6	9	8	3	0
14	5	6	9	2	3	2	30	5	6	9	4	3	2	46	5	6	9	6	3	2	62	5	6	9	8	3	2
15	5	6	9	2	3	4	31	5	6	9	4	3	4	47	5	6	9	6	3	4	63	5	6	9	8	3	4
16	5	6	9	2	3	6	32	5	6	9	4	3	6	48	5	6	9	6	3	6	64	5	6	9	8	3	6

HORIZONTAL MASTER (rows 1–4, 5–8, 9–12, 13–16)

BLOCK MASTER (below each block of four rows in each column)

3	2	1	0	5	8
5	4	3	2	7	0
7	6	5	4	9	2
9	8	7	6	1	4
1	0	9	8	3	6
f	e	d	c	b	a

THEORETICAL MASTER

BASE FIELD (KBA) KEY BITTING ARRAY

⇐ SEQUENCE OF PROGRESSION

PAGE MASTER ________

GROUP MASTER ________

SECTION MASTER ________

PAGE # 9 OF 64 PAGES

LINE MASTERS ⇩

VERTICAL MASTER / VERTICAL MASTER / VERTICAL MASTER / VERTICAL MASTER

#	f	e	d	c	b	a	#	f	e	d	c	b	a	#	f	e	d	c	b	a	#	f	e	d	c	b	a
1	5	8	3	2	7	0	17	5	8	3	4	7	0	33	5	8	3	6	7	0	49	5	8	3	8	7	0
2	5	8	3	2	7	2	18	5	8	3	4	7	2	34	5	8	3	6	7	2	50	5	8	3	8	7	2
3	5	8	3	2	7	4	19	5	8	3	4	7	4	35	5	8	3	6	7	4	51	5	8	3	8	7	4
4	5	8	3	2	7	6	20	5	8	3	4	7	6	36	5	8	3	6	7	6	52	5	8	3	8	7	6
5	5	8	3	2	9	0	21	5	8	3	4	9	0	37	5	8	3	6	9	0	53	5	8	3	8	9	0
6	5	8	3	2	9	2	22	5	8	3	4	9	2	38	5	8	3	6	9	2	54	5	8	3	8	9	2
7	5	8	3	2	9	4	23	5	8	3	4	9	4	39	5	8	3	6	9	4	55	5	8	3	8	9	4
8	5	8	3	2	9	6	24	5	8	3	4	9	6	40	5	8	3	6	9	6	56	5	8	3	8	9	6
9	5	8	3	2	1	0	25	5	8	3	4	1	0	41	5	8	3	6	1	0	57	5	8	3	8	1	0
10	5	8	3	2	1	2	26	5	8	3	4	1	2	42	5	8	3	6	1	2	58	5	8	3	8	1	2
11	5	8	3	2	1	4	27	5	8	3	4	1	4	43	5	8	3	6	1	4	59	5	8	3	8	1	4
12	5	8	3	2	1	6	28	5	8	3	4	1	6	44	5	8	3	6	1	6	60	5	8	3	8	1	6
13	5	8	3	2	3	0	29	5	8	3	4	3	0	45	5	8	3	6	3	0	61	5	8	3	8	3	0
14	5	8	3	2	3	2	30	5	8	3	4	3	2	46	5	8	3	6	3	2	62	5	8	3	8	3	2
15	5	8	3	2	3	4	31	5	8	3	4	3	4	47	5	8	3	6	3	4	63	5	8	3	8	3	4
16	5	8	3	2	3	6	32	5	8	3	4	3	6	48	5	8	3	6	3	6	64	5	8	3	8	3	6

HORIZONTAL MASTER (rows 1–4, 5–8, 9–12, 13–16)

BLOCK MASTER (below each block of four rows in each column)

3	2	1	0	5	8
5	4	3	2	7	0
7	6	5	4	9	2
9	8	7	6	1	4
1	0	9	8	3	6
f	e	d	c	b	a

THEORETICAL MASTER

BASE FIELD (KBA) KEY BITTING ARRAY

⇐ SEQUENCE OF PROGRESSION

PAGE MASTER

GROUP MASTER

SECTION MASTER

PAGE # 10 OF 64 PAGES

LINE MASTERS ⇩

	VERTICAL MASTER		VERTICAL MASTER		VERTICAL MASTER		VERTICAL MASTER
1	5 8 5 2 7 0	17	5 8 5 4 7 0	33	5 8 5 6 7 0	49	5 8 5 8 7 0
2	5 8 5 2 7 2	18	5 8 5 4 7 2	34	5 8 5 6 7 2	50	5 8 5 8 7 2
3	5 8 5 2 7 4	19	5 8 5 4 7 4	35	5 8 5 6 7 4	51	5 8 5 8 7 4
4	5 8 5 2 7 6	20	5 8 5 4 7 6	36	5 8 5 6 7 6	52	5 8 5 8 7 6
	BLOCK MASTER		BLOCK MASTER		BLOCK MASTER		BLOCK MASTER
5	5 8 5 2 9 0	21	5 8 5 4 9 0	37	5 8 5 6 9 0	53	5 8 5 8 9 0
6	5 8 5 2 9 2	22	5 8 5 4 9 2	38	5 8 5 6 9 2	54	5 8 5 8 9 2
7	5 8 5 2 9 4	23	5 8 5 4 9 4	39	5 8 5 6 9 4	55	5 8 5 8 9 4
8	5 8 5 2 9 6	24	5 8 5 4 9 6	40	5 8 5 6 9 6	56	5 8 5 8 9 6
	BLOCK MASTER		BLOCK MASTER		BLOCK MASTER		BLOCK MASTER
9	5 8 5 2 1 0	25	5 8 5 4 1 0	41	5 8 5 6 1 0	57	5 8 5 8 1 0
10	5 8 5 2 1 2	26	5 8 5 4 1 2	42	5 8 5 6 1 2	58	5 8 5 8 1 2
11	5 8 5 2 1 4	27	5 8 5 4 1 4	43	5 8 5 6 1 4	59	5 8 5 8 1 4
12	5 8 5 2 1 6	28	5 8 5 4 1 6	44	5 8 5 6 1 6	60	5 8 5 8 1 6
	BLOCK MASTER		BLOCK MASTER		BLOCK MASTER		BLOCK MASTER
13	5 8 5 2 3 0	29	5 8 5 4 3 0	45	5 8 5 6 3 0	61	5 8 5 8 3 0
14	5 8 5 2 3 2	30	5 8 5 4 3 2	46	5 8 5 6 3 2	62	5 8 5 8 3 2
15	5 8 5 2 3 4	31	5 8 5 4 3 4	47	5 8 5 6 3 4	63	5 8 5 8 3 4
16	5 8 5 2 3 6	32	5 8 5 4 3 6	48	5 8 5 6 3 6	64	5 8 5 8 3 6
	BLOCK MASTER		BLOCK MASTER		BLOCK MASTER		BLOCK MASTER

HORIZONTAL MASTER (one per block of four rows)

3	2	1	0	5	8
5	4	3	2	7	0
7	6	5	4	9	2
9	8	7	6	1	4
1	0	9	8	3	6
f	e	d	c	b	a

THEORETICAL MASTER

BASE FIELD (KBA) KEY BITTING ARRAY

⇐ SEQUENCE OF PROGRESSION

PAGE MASTER

GROUP MASTER

SECTION MASTER

PAGE # 11 OF 64 PAGES

LINE MASTERS ⇩

	VERTICAL MASTER		VERTICAL MASTER		VERTICAL MASTER		VERTICAL MASTER
1	5 8 7 2 7 0	17	5 8 7 4 7 0	33	5 8 7 6 7 0	49	5 8 7 8 7 0
2	5 8 7 2 7 2	18	5 8 7 4 7 2	34	5 8 7 6 7 2	50	5 8 7 8 7 2
3	5 8 7 2 7 4	19	5 8 7 4 7 4	35	5 8 7 6 7 4	51	5 8 7 8 7 4
4	5 8 7 2 7 6	20	5 8 7 4 7 6	36	5 8 7 6 7 6	52	5 8 7 8 7 6
	BLOCK MASTER		BLOCK MASTER		BLOCK MASTER		BLOCK MASTER
5	5 8 7 2 9 0	21	5 8 7 4 9 0	37	5 8 7 6 9 0	53	5 8 7 8 9 0
6	5 8 7 2 9 2	22	5 8 7 4 9 2	38	5 8 7 6 9 2	54	5 8 7 8 9 2
7	5 8 7 2 9 4	23	5 8 7 4 9 4	39	5 8 7 6 9 4	55	5 8 7 8 9 4
8	5 8 7 2 9 6	24	5 8 7 4 9 6	40	5 8 7 6 9 6	56	5 8 7 8 9 6
	BLOCK MASTER		BLOCK MASTER		BLOCK MASTER		BLOCK MASTER
9	5 8 7 2 1 0	25	5 8 7 4 1 0	41	5 8 7 6 1 0	57	5 8 7 8 1 0
10	5 8 7 2 1 2	26	5 8 7 4 1 2	42	5 8 7 6 1 2	58	5 8 7 8 1 2
11	5 8 7 2 1 4	27	5 8 7 4 1 4	43	5 8 7 6 1 4	59	5 8 7 8 1 4
12	5 8 7 2 1 6	28	5 8 7 4 1 6	44	5 8 7 6 1 6	60	5 8 7 8 1 6
	BLOCK MASTER		BLOCK MASTER		BLOCK MASTER		BLOCK MASTER
13	5 8 7 2 3 0	29	5 8 7 4 3 0	45	5 8 7 6 3 0	61	5 8 7 8 3 0
14	5 8 7 2 3 2	30	5 8 7 4 3 2	46	5 8 7 6 3 2	62	5 8 7 8 3 2
15	5 8 7 2 3 4	31	5 8 7 4 3 4	47	5 8 7 6 3 4	63	5 8 7 8 3 4
16	5 8 7 2 3 6	32	5 8 7 4 3 6	48	5 8 7 6 3 6	64	5 8 7 8 3 6
	BLOCK MASTER		BLOCK MASTER		BLOCK MASTER		BLOCK MASTER

HORIZONTAL MASTER (one per block of four rows)

3	2	1	0	5	8
5	4	3	2	7	0
7	6	5	4	9	2
9	8	7	6	1	4
1	0	9	8	3	6
f	e	d	c	b	a

THEORETICAL MASTER

BASE FIELD (KBA) KEY BITTING ARRAY

= SEQUENCE OF PROGRESSION

PAGE MASTER ______

GROUP MASTER ______

SECTION MASTER ______

PAGE # 12 OF 64 PAGES

LINE MASTERS ⇩

VERTICAL MASTER / HORIZONTAL MASTER / BLOCK MASTER

1	5	8	9	2	7	0	17	5	8	9	4	7	0	33	5	8	9	6	7	0	49	5	8	9	8	7	0
2	5	8	9	2	7	2	18	5	8	9	4	7	2	34	5	8	9	6	7	2	50	5	8	9	8	7	2
3	5	8	9	2	7	4	19	5	8	9	4	7	4	35	5	8	9	6	7	4	51	5	8	9	8	7	4
4	5	8	9	2	7	6	20	5	8	9	4	7	6	36	5	8	9	6	7	6	52	5	8	9	8	7	6
5	5	8	9	2	9	0	21	5	8	9	4	9	0	37	5	8	9	6	9	0	53	5	8	9	8	9	0
6	5	8	9	2	9	2	22	5	8	9	4	9	2	38	5	8	9	6	9	2	54	5	8	9	8	9	2
7	5	8	9	2	9	4	23	5	8	9	4	9	4	39	5	8	9	6	9	4	55	5	8	9	8	9	4
8	5	8	9	2	9	6	24	5	8	9	4	9	6	40	5	8	9	6	9	6	56	5	8	9	8	9	6
9	5	8	9	2	1	0	25	5	8	9	4	1	0	41	5	8	9	6	1	0	57	5	8	9	8	1	0
10	5	8	9	2	1	2	26	5	8	9	4	1	2	42	5	8	9	6	1	2	58	5	8	9	8	1	2
11	5	8	9	2	1	4	27	5	8	9	4	1	4	43	5	8	9	6	1	4	59	5	8	9	8	1	4
12	5	8	9	2	1	6	28	5	8	9	4	1	6	44	5	8	9	6	1	6	60	5	8	9	8	1	6
13	5	8	9	2	3	0	29	5	8	9	4	3	0	45	5	8	9	6	3	0	61	5	8	9	8	3	0
14	5	8	9	2	3	2	30	5	8	9	4	3	2	46	5	8	9	6	3	2	62	5	8	9	8	3	2
15	5	8	9	2	3	4	31	5	8	9	4	3	4	47	5	8	9	6	3	4	63	5	8	9	8	3	4
16	5	8	9	2	3	6	32	5	8	9	4	3	6	48	5	8	9	6	3	6	64	5	8	9	8	3	6

3	2	1	0	5	8
5	4	3	2	7	0
7	6	5	4	9	2
9	8	7	6	1	4
1	0	9	8	3	6
f	e	d	c	b	a

THEORETICAL MASTER

BASE FIELD (KBA) KEY BITTING ARRAY

= SEQUENCE OF PROGRESSION

PAGE MASTER ______

GROUP MASTER ______

SECTION MASTER ______

PAGE # 13 OF 64 PAGES

LINE MASTERS ⇩

VERTICAL MASTER / HORIZONTAL MASTER / BLOCK MASTER

1	5	0	3	2	7	0	17	5	0	3	4	7	0	33	5	0	3	6	7	0	49	5	0	3	8	7	0
2	5	0	3	2	7	2	18	5	0	3	4	7	2	34	5	0	3	6	7	2	50	5	0	3	8	7	2
3	5	0	3	2	7	4	19	5	0	3	4	7	4	35	5	0	3	6	7	4	51	5	0	3	8	7	4
4	5	0	3	2	7	6	20	5	0	3	4	7	6	36	5	0	3	6	7	6	52	5	0	3	8	7	6
5	5	0	3	2	9	0	21	5	0	3	4	9	0	37	5	0	3	6	9	0	53	5	0	3	8	9	0
6	5	0	3	2	9	2	22	5	0	3	4	9	2	38	5	0	3	6	9	2	54	5	0	3	8	9	2
7	5	0	3	2	9	4	23	5	0	3	4	9	4	39	5	0	3	6	9	4	55	5	0	3	8	9	4
8	5	0	3	2	9	6	24	5	0	3	4	9	6	40	5	0	3	6	9	6	56	5	0	3	8	9	6
9	5	0	3	2	1	0	25	5	0	3	4	1	0	41	5	0	3	6	1	0	57	5	0	3	8	1	0
10	5	0	3	2	1	2	26	5	0	3	4	1	2	42	5	0	3	6	1	2	58	5	0	3	8	1	2
11	5	0	3	2	1	4	27	5	0	3	4	1	4	43	5	0	3	6	1	4	59	5	0	3	8	1	4
12	5	0	3	2	1	6	28	5	0	3	4	1	6	44	5	0	3	6	1	6	60	5	0	3	8	1	6
13	5	0	3	2	3	0	29	5	0	3	4	3	0	45	5	0	3	6	3	0	61	5	0	3	8	3	0
14	5	0	3	2	3	2	30	5	0	3	4	3	2	46	5	0	3	6	3	2	62	5	0	3	8	3	2
15	5	0	3	2	3	4	31	5	0	3	4	3	4	47	5	0	3	6	3	4	63	5	0	3	8	3	4
16	5	0	3	2	3	6	32	5	0	3	4	3	6	48	5	0	3	6	3	6	64	5	0	3	8	3	6

3	2	1	0	5	8
5	4	3	2	7	0
7	6	5	4	9	2
9	8	7	6	1	4
1	0	9	8	3	6
f	e	d	c	b	a

THEORETICAL MASTER

BASE FIELD (KBA) KEY BITTING ARRAY

⇐ SEQUENCE OF PROGRESSION

PAGE MASTER ______

GROUP MASTER ______

SECTION MASTER ______

PAGE # 14 OF 64 PAGES

LINE MASTERS ⇩

VERTICAL MASTER | VERTICAL MASTER | VERTICAL MASTER | VERTICAL MASTER

HORIZONTAL MASTER

1	5	0	5	2	7	0	17	5	0	5	4	7	0	33	5	0	5	6	7	0	49	5	0	5	8	7	0
2	5	0	5	2	7	2	18	5	0	5	4	7	2	34	5	0	5	6	7	2	50	5	0	5	8	7	2
3	5	0	5	2	7	4	19	5	0	5	4	7	4	35	5	0	5	6	7	4	51	5	0	5	8	7	4
4	5	0	5	2	7	6	20	5	0	5	4	7	6	36	5	0	5	6	7	6	52	5	0	5	8	7	6

BLOCK MASTER | BLOCK MASTER | BLOCK MASTER | BLOCK MASTER

HORIZONTAL MASTER

5	5	0	5	2	9	0	21	5	0	5	4	9	0	37	5	0	5	6	9	0	53	5	0	5	8	9	0
6	5	0	5	2	9	2	22	5	0	5	4	9	2	38	5	0	5	6	9	2	54	5	0	5	8	9	2
7	5	0	5	2	9	4	23	5	0	5	4	9	4	39	5	0	5	6	9	4	55	5	0	5	8	9	4
8	5	0	5	2	9	6	24	5	0	5	4	9	6	40	5	0	5	6	9	6	56	5	0	5	8	9	6

BLOCK MASTER | BLOCK MASTER | BLOCK MASTER | BLOCK MASTER

HORIZONTAL MASTER

9	5	0	5	2	1	0	25	5	0	5	4	1	0	41	5	0	5	6	1	0	57	5	0	5	8	1	0
10	5	0	5	2	1	2	26	5	0	5	4	1	2	42	5	0	5	6	1	2	58	5	0	5	8	1	2
11	5	0	5	2	1	4	27	5	0	5	4	1	4	43	5	0	5	6	1	4	59	5	0	5	8	1	4
12	5	0	5	2	1	6	28	5	0	5	4	1	6	44	5	0	5	6	1	6	60	5	0	5	8	1	6

BLOCK MASTER | BLOCK MASTER | BLOCK MASTER | BLOCK MASTER

HORIZONTAL MASTER

13	5	0	5	2	3	0	29	5	0	5	4	3	0	45	5	0	5	6	3	0	61	5	0	5	8	3	0
14	5	0	5	2	3	2	30	5	0	5	4	3	2	46	5	0	5	6	3	2	62	5	0	5	8	3	2
15	5	0	5	2	3	4	31	5	0	5	4	3	4	47	5	0	5	6	3	4	63	5	0	5	8	3	4
16	5	0	5	2	3	6	32	5	0	5	4	3	6	48	5	0	5	6	3	6	64	5	0	5	8	3	6

BLOCK MASTER | BLOCK MASTER | BLOCK MASTER | BLOCK MASTER

3	2	1	0	5	8
5	4	3	2	7	0
7	6	5	4	9	2
9	8	7	6	1	4
1	0	9	8	3	6
f	e	d	c	b	a

THEORETICAL MASTER

BASE FIELD (KBA) KEY BITTING ARRAY

⇐ SEQUENCE OF PROGRESSION

PAGE MASTER ______

GROUP MASTER ______

SECTION MASTER ______

PAGE # 15 OF 64 PAGES

LINE MASTERS ⇩

VERTICAL MASTER | VERTICAL MASTER | VERTICAL MASTER | VERTICAL MASTER

HORIZONTAL MASTER

1	5	0	7	2	7	0	17	5	0	7	4	7	0	33	5	0	7	6	7	0	49	5	0	7	8	7	0
2	5	0	7	2	7	2	18	5	0	7	4	7	2	34	5	0	7	6	7	2	50	5	0	7	8	7	2
3	5	0	7	2	7	4	19	5	0	7	4	7	4	35	5	0	7	6	7	4	51	5	0	7	8	7	4
4	5	0	7	2	7	6	20	5	0	7	4	7	6	36	5	0	7	6	7	6	52	5	0	7	8	7	6

BLOCK MASTER | BLOCK MASTER | BLOCK MASTER | BLOCK MASTER

HORIZONTAL MASTER

5	5	0	7	2	9	0	21	5	0	7	4	9	0	37	5	0	7	6	9	0	53	5	0	7	8	9	0
6	5	0	7	2	9	2	22	5	0	7	4	9	2	38	5	0	7	6	9	2	54	5	0	7	8	9	2
7	5	0	7	2	9	4	23	5	0	7	4	9	4	39	5	0	7	6	9	4	55	5	0	7	8	9	4
8	5	0	7	2	9	6	24	5	0	7	4	9	6	40	5	0	7	6	9	6	56	5	0	7	8	9	6

BLOCK MASTER | BLOCK MASTER | BLOCK MASTER | BLOCK MASTER

HORIZONTAL MASTER

9	5	0	7	2	1	0	25	5	0	7	4	1	0	41	5	0	7	6	1	0	57	5	0	7	8	1	0
10	5	0	7	2	1	2	26	5	0	7	4	1	2	42	5	0	7	6	1	2	58	5	0	7	8	1	2
11	5	0	7	2	1	4	27	5	0	7	4	1	4	43	5	0	7	6	1	4	59	5	0	7	8	1	4
12	5	0	7	2	1	6	28	5	0	7	4	1	6	44	5	0	7	6	1	6	60	5	0	7	8	1	6

BLOCK MASTER | BLOCK MASTER | BLOCK MASTER | BLOCK MASTER

HORIZONTAL MASTER

13	5	0	7	2	3	0	29	5	0	7	4	3	0	45	5	0	7	6	3	0	61	5	0	7	8	3	0
14	5	0	7	2	3	2	30	5	0	7	4	3	2	46	5	0	7	6	3	2	62	5	0	7	8	3	2
15	5	0	7	2	3	4	31	5	0	7	4	3	4	47	5	0	7	6	3	4	63	5	0	7	8	3	4
16	5	0	7	2	3	6	32	5	0	7	4	3	6	48	5	0	7	6	3	6	64	5	0	7	8	3	6

BLOCK MASTER | BLOCK MASTER | BLOCK MASTER | BLOCK MASTER

	3	2	1	0	5	8
	5	4	3	2	7	0
	7	6	5	4	9	2
	9	8	7	6	1	4
	1	0	9	8	3	6
	f	e	d	c	b	a

THEORETICAL MASTER

BASE FIELD (KBA) KEY BITTING ARRAY

⇐ SEQUENCE OF PROGRESSION

PAGE MASTER ______

GROUP MASTER ______

SECTION MASTER ______

PAGE # 16 OF 64 PAGES

LINE MASTERS ⇩

	VERTICAL MASTER		VERTICAL MASTER		VERTICAL MASTER		VERTICAL MASTER	LINE MASTERS
1	5 0 9 2 7 0	17	5 0 9 4 7 0	33	5 0 9 6 7 0	49	5 0 9 8 7 0	
2	5 0 9 2 7 2	18	5 0 9 4 7 2	34	5 0 9 6 7 2	50	5 0 9 8 7 2	
HORIZONTAL MASTER 3	5 0 9 2 7 4	19	5 0 9 4 7 4	35	5 0 9 6 7 4	51	5 0 9 8 7 4	
4	5 0 9 2 7 6	20	5 0 9 4 7 6	36	5 0 9 6 7 6	52	5 0 9 8 7 6	
	BLOCK MASTER		BLOCK MASTER		BLOCK MASTER		BLOCK MASTER	
5	5 0 9 2 9 0	21	5 0 9 4 9 0	37	5 0 9 6 9 0	53	5 0 9 8 9 0	
6	5 0 9 2 9 2	22	5 0 9 4 9 2	38	5 0 9 6 9 2	54	5 0 9 8 9 2	
HORIZONTAL MASTER 7	5 0 9 2 9 4	23	5 0 9 4 9 4	39	5 0 9 6 9 4	55	5 0 9 8 9 4	
8	5 0 9 2 9 6	24	5 0 9 4 9 6	40	5 0 9 6 9 6	56	5 0 9 8 9 6	
	BLOCK MASTER		BLOCK MASTER		BLOCK MASTER		BLOCK MASTER	
9	5 0 9 2 1 0	25	5 0 9 4 1 0	41	5 0 9 6 1 0	57	5 0 9 8 1 0	
10	5 0 9 2 1 2	26	5 0 9 4 1 2	42	5 0 9 6 1 2	58	5 0 9 8 1 2	
HORIZONTAL MASTER 11	5 0 9 2 1 4	27	5 0 9 4 1 4	43	5 0 9 6 1 4	59	5 0 9 8 1 4	
12	5 0 9 2 1 6	28	5 0 9 4 1 6	44	5 0 9 6 1 6	60	5 0 9 8 1 6	
	BLOCK MASTER		BLOCK MASTER		BLOCK MASTER		BLOCK MASTER	
13	5 0 9 2 3 0	29	5 0 9 4 3 0	45	5 0 9 6 3 0	61	5 0 9 8 3 0	
14	5 0 9 2 3 2	30	5 0 9 4 3 2	46	5 0 9 6 3 2	62	5 0 9 8 3 2	
HORIZONTAL MASTER 15	5 0 9 2 3 4	31	5 0 9 4 3 4	47	5 0 9 6 3 4	63	5 0 9 8 3 4	
16	5 0 9 2 3 6	32	5 0 9 4 3 6	48	5 0 9 6 3 6	64	5 0 9 8 3 6	
	BLOCK MASTER		BLOCK MASTER		BLOCK MASTER		BLOCK MASTER	

	3	2	1	0	5	8
	5	4	3	2	7	0
	7	6	5	4	9	2
	9	8	7	6	1	4
	1	0	9	8	3	6
	f	e	d	c	b	a

THEORETICAL MASTER

BASE FIELD (KBA) KEY BITTING ARRAY

⇐ SEQUENCE OF PROGRESSION

PAGE MASTER ______

GROUP MASTER ______

SECTION MASTER ______

PAGE # 17 OF 64 PAGES

LINE MASTERS ⇩

	VERTICAL MASTER		VERTICAL MASTER		VERTICAL MASTER		VERTICAL MASTER	LINE MASTERS
1	7 4 3 2 7 0	17	7 4 3 4 7 0	33	7 4 3 6 7 0	49	7 4 3 8 7 0	
2	7 4 3 2 7 2	18	7 4 3 4 7 2	34	7 4 3 6 7 2	50	7 4 3 8 7 2	
HORIZONTAL MASTER 3	7 4 3 2 7 4	19	7 4 3 4 7 4	35	7 4 3 6 7 4	51	7 4 3 8 7 4	
4	7 4 3 2 7 6	20	7 4 3 4 7 6	36	7 4 3 6 7 6	52	7 4 3 8 7 6	
	BLOCK MASTER		BLOCK MASTER		BLOCK MASTER		BLOCK MASTER	
5	7 4 3 2 9 0	21	7 4 3 4 9 0	37	7 4 3 6 9 0	53	7 4 3 8 9 0	
6	7 4 3 2 9 2	22	7 4 3 4 9 2	38	7 4 3 6 9 2	54	7 4 3 8 9 2	
HORIZONTAL MASTER 7	7 4 3 2 9 4	23	7 4 3 4 9 4	39	7 4 3 6 9 4	55	7 4 3 8 9 4	
8	7 4 3 2 9 6	24	7 4 3 4 9 6	40	7 4 3 6 9 6	56	7 4 3 8 9 6	
	BLOCK MASTER		BLOCK MASTER		BLOCK MASTER		BLOCK MASTER	
9	7 4 3 2 1 0	25	7 4 3 4 1 0	41	7 4 3 6 1 0	57	7 4 3 8 1 0	
10	7 4 3 2 1 2	26	7 4 3 4 1 2	42	7 4 3 6 1 2	58	7 4 3 8 1 2	
HORIZONTAL MASTER 11	7 4 3 2 1 4	27	7 4 3 4 1 4	43	7 4 3 6 1 4	59	7 4 3 8 1 4	
12	7 4 3 2 1 6	28	7 4 3 4 1 6	44	7 4 3 6 1 6	60	7 4 3 8 1 6	
	BLOCK MASTER		BLOCK MASTER		BLOCK MASTER		BLOCK MASTER	
13	7 4 3 2 3 0	29	7 4 3 4 3 0	45	7 4 3 6 3 0	61	7 4 3 8 3 0	
14	7 4 3 2 3 2	30	7 4 3 4 3 2	46	7 4 3 6 3 2	62	7 4 3 8 3 2	
HORIZONTAL MASTER 15	7 4 3 2 3 4	31	7 4 3 4 3 4	47	7 4 3 6 3 4	63	7 4 3 8 3 4	
16	7 4 3 2 3 6	32	7 4 3 4 3 6	48	7 4 3 6 3 6	64	7 4 3 8 3 6	
	BLOCK MASTER		BLOCK MASTER		BLOCK MASTER		BLOCK MASTER	

Now that we have finished a chart (or at least 17 of the 64 pages of it) we can look back and see how the two page SPF shortcut works.

Let's pick a page at random from the 17 above. For now we will choose page 9. On the top chart of the two page SPF shortcut (on book page 126 – the next page) locate the small 9 next to the bittings. It is the top bitting in the third block down and the cuts next to it are 5 8 3.

Now look at page 9 from the charts we just did, and you can see that the entire page has as its f, e and d position cuts the 5, 8 and 3. The remaining cuts are those shown in the bottom chart of the shorthand SPF on book page 129 (the next page).

Try it with any of the pages, and you will find it works the same way. That is how the shortcut can be used instead of having to complete all the pages of the system.

Two Page Shortcut

3	2	1	0	5	8
5	4	3	2	7	0
7	6	5	4	9	2
9	8	7	6	1	4
1	0	9	8	3	6
f	e	d			

THEORETICAL MASTER

BASE FIELD (KBA) KEY BITTING ARRAY

⇦ SEQUENCE OF PROGRESSION

PAGE MASTER ____ GROUP MASTER ____ SECTION MASTER ____

PAGE # ____ OF ____ PAGES

LINE MASTERS ⇩

VERTICAL MASTER | VERTICAL MASTER | VERTICAL MASTER | VERTICAL MASTER

HORIZONTAL MASTER

1	5	4	3				17	7	4	3				33	9	4	3				49	1	4	3			
2	5	4	5				18	7	4	5				34	9	4	5				50	1	4	5			
3	5	4	7				19	7	4	7				35	9	4	7				51	1	4	7			
4	5	4	9				20	7	4	9				36	9	4	9				52	1	4	9			

BLOCK MASTER | BLOCK MASTER | BLOCK MASTER | BLOCK MASTER

HORIZONTAL MASTER

5	5	6	3				21	7	6	3				37	9	6	3				53	1	6	3			
6	5	6	5				22	7	6	5				38	9	6	5				54	1	6	5			
7	5	6	7				23	7	6	7				39	9	6	7				55	1	6	7			
8	5	6	9				24	7	6	9				40	9	6	9				56	1	6	9			

BLOCK MASTER | BLOCK MASTER | BLOCK MASTER | BLOCK MASTER

HORIZONTAL MASTER

9	5	8	3				25	7	8	3				41	9	8	3				57	1	8	3			
10	5	8	5				26	7	8	5				42	9	8	5				58	1	8	5			
11	5	8	7				27	7	8	7				43	9	8	7				59	1	8	7			
12	5	8	9				28	7	8	9				44	9	8	9				60	1	8	9			

BLOCK MASTER | BLOCK MASTER | BLOCK MASTER | BLOCK MASTER

HORIZONTAL MASTER

13	5	0	3				29	7	0	3				45	9	0	3				61	1	0	3			
14	5	0	5				30	7	0	5				46	9	0	5				62	1	0	5			
15	5	0	7				31	7	0	7				47	9	0	7				63	1	0	7			
16	5	0	9				32	7	0	9				48	9	0	9				64	1	0	9			

BLOCK MASTER | BLOCK MASTER | BLOCK MASTER | BLOCK MASTER

Together, these two charts represent 64 pages.

3	2	1	0	5	8
5	4	3	2	7	0
7	6	5	4	9	2
9	8	7	6	1	4
1	0	9	8	3	6
f	e	d	c	b	a

THEORETICAL MASTER

BASE FIELD (KBA) KEY BITTING ARRAY

⇦ SEQUENCE OF PROGRESSION

PAGE MASTER ____ GROUP MASTER ____ SECTION MASTER ____

PAGE # ____ OF ____ PAGES

LINE MASTERS ⇩

VERTICAL MASTER | VERTICAL MASTER | VERTICAL MASTER | VERTICAL MASTER

HORIZONTAL MASTER

1				2	7	0	17				4	7	0	33				6	7	0	49				8	7	0
2				2	7	2	18				4	7	2	34				6	7	2	50				8	7	2
3				2	7	4	19				4	7	4	35				6	7	4	51				8	7	4
4				2	7	6	20				4	7	6	36				6	7	6	52				8	7	6

BLOCK MASTER | BLOCK MASTER | BLOCK MASTER | BLOCK MASTER

HORIZONTAL MASTER

5				2	9	0	21				4	9	0	37				6	9	0	53				8	9	0
6				2	9	2	22				4	9	2	38				6	9	2	54				8	9	2
7				2	9	4	23				4	9	4	39				6	9	4	55				8	9	4
8				2	9	6	24				4	9	6	40				6	9	6	56				8	9	6

BLOCK MASTER | BLOCK MASTER | BLOCK MASTER | BLOCK MASTER

HORIZONTAL MASTER

9				2	1	0	25				4	1	0	41				6	1	0	57				8	1	0
10				2	1	2	26				4	1	2	42				6	1	2	58				8	1	2
11				2	1	4	27				4	1	4	43				6	1	4	59				8	1	4
12				2	1	6	28				4	1	6	44				6	1	6	60				8	1	6

BLOCK MASTER | BLOCK MASTER | BLOCK MASTER | BLOCK MASTER

HORIZONTAL MASTER

13				2	3	0	29				4	3	0	45				6	3	0	61				8	3	0
14				2	3	2	30				4	3	2	46				6	3	2	62				8	3	2
15				2	3	4	31				4	3	4	47				6	3	4	63				8	3	4
16				2	3	6	32				4	3	6	48				6	3	6	64				8	3	6

BLOCK MASTER | BLOCK MASTER | BLOCK MASTER | BLOCK MASTER

LESSON SEVEN
ADDING MASTER KEY LEVELS
(BREAKING THE PROGRESSION INTO SYSTEMS)

So far, we have dealt with possible change keys based on the TMK, the Top (Theoretical) Master Key.

But what if our customer's system had a need for multiple levels of master key? The good news is that our SPF has more on it than we have used so far, and what has been missing is very useful, addressing the need for multiple levels. Using an example of a TMK of 334104, let's look at what happens when we pin cylinders to the change key and TMK.

With a Top (Theoretical) Master of 334104, the first four change keys would be 110320, 110322, 110326, 110328.

Using the basic pinning rules from the section on Master Pinning, we see that we would compare the two keys we want to work in the cylinder (in this case the TMK and the change key). The smallest of the cuts in each column is used as the bottom pin, and the difference between the two cuts in each column is the master pin size used.

Thus:	
TMK:	334104
Change Key	110320
Smallest = Bottom Pins	110100
Difference = Master Pins	224224

Now, both keys work! Isn't that wonderful?

BUT NOT SO WONDERFUL is the fact that each of the following keys also would work it IF they existed. They are called **INCIDENTAL MASTER KEYS** or **Incidental Keys**. They are often historically referred to as "**Ghost**" keys or "**Phantom**" keys, but either term will draw a lot of criticism today, having completely fallen into disuse and ill repute.

334100	330100	134100	130100	314100	310100	114100	110100
334104	330104	134104	130104	314104	314104	114104	110104
334120	330120	134120	130120	314120	310120	114120	110120
334124	330124	134124	130124	314124	310124	114124	110124
334300	330300	134300	130300	314300	310300	114300	110300
334304	330304	134304	130304	314304	310304	114304	110304
334320	330320	134320	130320	314320	310320	114320	110320
334324	330324	134324	130324	314324	310324	114324	110324

Related Terms from the LIST Council Dictionary:

incidental master key
n. a key cut to an unplanned shear line created when the cylinder is combinated to the top master key and a change key

That's 62 keys we DIDN'T intend to work it that WILL! The good news is that we can assign many of them "jobs" within the system to accomplish the needs of the customer's system.

Let's look at the first group of locks in our 17 page Sample progression chart. They were

1	5	4	3	2	7	0
2	5	4	3	2	7	2
3	5	4	3	2	7	4
4	5	4	3	2	7	6

BLOCK MASTER

Pinning them to the Theoretical Master would give us:

Theoretical Master	321058	321058	321058	321058
Change Key	543270	543272	543274	543276
Bottom Pins	321050	321052	321054	321056
Master Pins	222222	222226	222224	222222

If we wanted a key that could control JUST these four locks, out of the whole system, it would have to be as similar to these four change keys as possible, and yet have something in common with the TMK as well.

Taking a look at the four change keys, it is apparent that all of them have 54327 for the first five cuts, and if we were to look at the rest of the system, we would see that they are the only four that do!

In the last chamber, however, one has a 0, one has a 2, one has a 4, and the other has an 6. ALL OF THEM CAN BE OPENED BY THE THEORETICAL MASTER, HOWEVER, WHICH HAS A # 8 CUT IN THAT POSITION! Thus a key that is cut to the pattern 5-4-3-2-7-8 would open all four of them, and no others in the system!

Because all four of the combinations it opened were in one BLOCK of the Change Key Progression Chart, for identification purposes we will call it a **Block Master**. Two other names it has been called are the **Level One Submaster**, and the **One Pin Master**, but these names have largely fallen into disuse.

Related Terms from the LIST Council Dictionary:

block master key
n. the one pin master key for all combinations listed as a block in the standard progression format cylinder which may be operated in a clockwise and counterclockwise direction by a single key

A space has been provided under the Block for each Block Master. We could do the same for every Block Master in the system – use the five cuts each change key has in common from Progression Sequences f, e, d, c and b, and in the a position (FIRST PROGRESSION SEQUENCE) use the cut from that position of our TMK.
It is that simple!

We could follow a similar pattern for a master capable of operating the 16 bittings (4 blocks) in a vertical column. Let's look at the first column:

VERTICAL MASTER

1	5	4	3	2	7	0
2	5	4	3	2	7	2
3	5	4	3	2	7	4
4	5	4	3	2	7	6

BLOCK MASTER

5	5	4	3	2	9	0
6	5	4	3	2	9	2
7	5	4	3	2	9	4
8	5	4	3	2	9	6

BLOCK MASTER

9	5	4	3	2	1	0
10	5	4	3	2	1	2
11	5	4	3	2	1	4
12	5	4	3	2	1	6

BLOCK MASTER

13	5	4	3	2	3	0
14	5	4	3	2	3	2
15	5	4	3	2	3	4
16	5	4	3	2	3	6

BLOCK MASTER

These four blocks all have 5-4-3-2-4- in the first five positions. Once again, they are all different in the far right position (which, in this case, is the first progression sequence), but are all able to be opened by the Theoretical Master, which in this case has a # 8 cut in that position.

But in the b (SECOND PROGRESSION SEQUENCE) position, they have 7, 9, 1 and 3. Therefore, because they differ in the b position as well as the a position, the master key for just these bittings, AND NO OTHER CHANGE KEY BITTINGS in the system, would need to use the cuts from the TMK in positions a and b. Therefore, for this example, it would be 5 4 3 2 5 8. The same pattern would continue for every Vertical Column Master Key in the system. A space has been provided above each column for us to list the VMK (Vertical Master Key) and in each case it will use the cuts that all the keys in the column have in common in positions f, e, d and c, and will use the cuts the TMK has in positions b and a.

Vertical Master Keys have also been called **Level Two Column Masters**, and **Two Pin Masters**, It is always in the benefit of a master keying Locksmith to recognize all the terms that have existed over the years, but it is important for you to use the approved terms.

Related Terms from the LIST Council Dictionary:

vertical group master key
n. the two-pin master key for all combinations listed in all blocks in a line down a page in the standard progression format

We can do this with HORIZONTAL (sideways) groups, also. Looking at the Blocks in a Horizontal row across the top of the first page of our Progression Chart from the Sample, we see the following combinations:

		VERTICAL MASTER							VERTICAL MASTER							VERTICAL MASTER							VERTICAL MASTER					
	1	5	4	3	2	7	0	17	5	4	3	4	7	0	33	5	4	3	6	7	0	49	5	4	3	8	7	0
	2	5	4	3	2	7	2	18	5	4	3	4	7	2	34	5	4	3	6	7	2	50	5	4	3	8	7	2
HORIZONTAL MASTER	3	5	4	3	2	7	4	19	5	4	3	4	7	4	35	5	4	3	6	7	4	51	5	4	3	8	7	4
	4	5	4	3	2	7	6	20	5	4	3	4	7	6	36	5	4	3	6	7	6	52	5	4	3	8	7	6

This time they all have a 5 4 3 in the first three positions from the left (position sequences f, e and d), and all have a number 7 cut in the second from the right (b position). In the fourth and sixth positions they are different, however. If all are opened by the TMK, it means that, in this example, the c position can be opened by a 0 (zero) cut, and the a (far right) position can be opened by an eight (8) cut.

Therefore, a key that is cut to 5 4 3 0 7 8 would open ALL these combinations, and no other locks in the system. This would be the **HORIZONTAL MASTER**.

The Horizontal Master always uses the cuts from the Theoretical Master for progression sequences a and c. The remaining cuts on the Horizontal Master are those that are common among all the change keys involved.

There is a space provided to the left of the Blocks involved for the Horizontal Master to be written in.

> **Related Terms from the LIST Council Dictionary:**
>
> **horizontal group master key**
> n. the two pin master key for all combinations listed in all blocks in a line across the page in the standard progression format

The same pattern can be used on every master key system using Total Position Progression to determine the Horizontal Master key. Every HMK uses the TMK cuts from positions a and c (regardless of where they are on the key) combined with the cuts from the rest of the change key positions in that horizontal group.

We could also find the Master key for a single line (or row) by comparing what the change keys in that line (or row) have in common and using the TMK cut where they have different cut depths.

For example, the first line on our chart looks like this:

1	5	4	3	2	7	0	17	5	4	3	4	7	0	33	5	4	3	6	7	0	49	5	4	3	8	7	0

By using only the c (THIRD progression sequence) position cut from the TMK, we can create a key that will control the combination in a straight line across the chart.

Therefore, the Line Master Key = 5 4 3 0 7 0

This is called the **LINE MASTER** and there is a space for one to the right of each line on the chart. .

The pattern would continue for every line on every page, to use only the c position (third progression) cut from the TMK, and with it to put the cuts common to that particular group of keys.

Space has been provided on the right of the charts to fill in the **Line Masters**. By the way, they have also been called "**Row Masters**", or "**One Pin Across Masters**".

> **Related Terms from the LIST Council Dictionary:**
>
> **row master key**
> n. the one-pin master key for all combinations listed on the same line across a page in the standard progression format

Now let's assume that we need a master capable of controlling an entire page, but no more than that. Looking at the page of our example, we see that each page in the system uses the same three cuts in the d, e and f positions throughout the page, but that what the three cuts are is different on every page.

We also see that variations exist throughout the a, b and c positions so that there is no capability to have a key that operates all of them unless it uses the TMK cuts for those positions.

Looking at our first page from the Sample, we see that the first three positions from the left are constant throughout the page, all having a cut pattern of 5-4-3- - -. This is because we used a right to left progression sequence and these are the positions, therefore, for the fourth, fifth, and sixth progression sequences (d, e and f).

Since all the locks in our system can be opened by our TMK, which has cuts at the tip of the key of - - -0-5-8, we can see that using the three cuts the page has in common, 5-4-3- - -, and combining them with the three cuts from the Theoretical Master where the keys differ from each other, we end up with a key of 5-4-3-0-5-8, which can control all the locks on the entire page, but no locks from any other page of the system.

The second page has the bitting pattern beginning with 5-4-5- - -, so the Master to control just that page of the system is 5-4-5-0-5-8.

The third page of this system, having 5-4-7- - -, would use a Master for the page of 5-4-7-0-5-8. And so on, throughout the system.

The master key to control an entire PAGE of the system, but no more, is generally referred to as a **PAGE MASTER**. It has also been called a "**Level 3 Segment Master**", and "**Three Pin Master**". Once again, it is to your benefit to recognize ALL the terms from all the teaching STYLES used by various master keying instructors and manuals, but to use the approved terms

Related Terms from the LIST Council Dictionary:

page master key
n. the three-pin master key for all combinations listed on a page in the standard progression format

.

The Page Masters ALWAYS use the cuts from the Theoretical Master for progression sequences a, b, and c, and the cuts all the keys have in common on the page from the d, e and f progression sequences. Remember that if we were varying the progression sequence it might not be the three closest to the bow.

THERE IS A SPACE PROVIDED FOR LISTING THE PAGE MASTER AT THE TOP RIGHT OF EACH PAGE.

All of the masters we have seen so far--BLOCK, VERTICAL, HORIZONTAL, LINE, and PAGE---are based on the same rules and terms regardless of the numbering system used by the manufacturer of the lock, or the KBA, although, as you saw in the section on Progression, the pages themselves vary widely.

However, the next level of control is affected by the base field for the fourth progression sequence. If it is "BASE TWO", (two numbers in the base field), it will control TWO pages of progression charts in the system. If it is "BASE THREE", (three numbers in the base field), it will control THREE pages of progression charts. If it is "BASE FOUR", (four numbers in each position of the KBA), it will control FOUR pages of progression. If it is BASE FIVE, (Five numbers in each position of the KBA), it will control FIVE pages of progression charts. etc.

Because it is the number of cut possibilities of the Fourth progression sequence that determines how many pages of progression charts it operates, it is more logical to note that it operates a GROUP OF PAGES, it has been called the "**GROUP MASTER**".

When it is in "BASE FOUR", it has been called the **Quad Master**, because it operates a quadrant of four pages. It has been referred to as a "**Level Four**" master, because it uses four of the cuts from the Theoretical Master, and likewise, it has been called a **Four Pin Master**. Pick a name you like, but learn to recognize the rest as well. But remember that these names make little or no sense when applied to a "Base Five" system or other KBA patterns.

Regardless of the name we call it, **it always uses the cuts from the Theoretical Master for progression sequences a, b, c and d, and the cuts that all the pages in the group have in common from the e and f progression sequences.**

In our example, since we were dealing with Right to Left Progression, the first such master would use the cuts 5-4---- from the common cuts, and ---1-0-5-8 from the TMK, giving us a master of 5 4 1 0 5 8, which, in our "Base four" system, would control the first four pages. 5 6 1 0 5 8 would control pages five through eight. 5 8 1 0 8, the next 4 pages , etc.

The last section we can break this system up into depends on the number of possibilities of the FIFTH PROGRESSION SEQUENCE. If there are two possibilities there, the master we now create will operate TWO GROUPS of pages. If there are three possibilities there, it will operate THREE GROUPS of pages. If there are four possibilities there, it will operate FOUR GROUPS of pages. If there are five possibilities there, it operates FIVE GROUPS of pages.

We refer to this as a **SECTION MASTER**, since it operates a large SECTION of our system. It has also been called a **"Level 5" master** and a **"5 Pin" master**.

This uses the 5 cuts from the TMK for cut positions a, b, c, d, e and puts them with the cut that the entire section has in common. In our example, our first section master would be 521058, and the section master for the second section would be 721058. The third section would be controlled by 921058, etc.

If we had a 5 pin key (and/or were pinning five chambers in the lock), the key, having only five cuts on it, would be the exact same for the theoretical master as for our section master, so it would not be capable of further division.

There is something we said in this lesson that should be further clarified. First, we have used the term "**OPERATES**" (etc.) to describe the action of the Master Key in relation to the change key. Most experts on the subject resent the commonly heard term "controls", and prefer the term "**OPERATES**".

Once the chart is filled in completely, we would cross out any unacceptable combinations that exceed the MACS. Since our Sample used a lock with possible depths of Zero to Nine, and we said a MACS of Seven was the most common, this would mean that we cross off any individual combination (whether master or change key) that operates a Zero next to a Nine, Zero next to an eight, or a One next to a Nine.

We cross them out no matter where they appear, even if they are master key combinations, BUT REMEMBER: WE CROSS OUT ONLY THE UNACCEPTABLE COMBINATION ITSELF, NOT ANY OTHER COMBINATIONS NEAR IT OR OPERATED BY IT!!!

If we cross out a Vertical Master, for example, that doesn't necessarily affect the combinations for change keys or Block Masters under it, or for other Vertical Masters next to it.

For example, it is not uncommon to have a need within a system for combinations for which there are no lower level master keys. What better place to pull the bittings from that a page where the page master and Vertical Master Keys are all unable to be accurately cut, but where the individual bittings are not?

Another purpose is when we need Single Keyed locks (for which no master key in the system will operate them). We reduce the possibilities to virtually nonexistent by first not pinning the master pins into the cylinder or core, and the fact that the lower level master keys that would have fallen above it cannot even be cut is, as the saying goes, the icing on the cake. How sweet it is!

But remember, you cannot use any key which cannot be accurately cut, so when crossing out, be certain you don't miss any keys that have undesirable combinations, especially those with MACS violations!

This being said, it is not uncommon to come across a key created by the manufacturer that violates the MACS and still works. But for how long will it continue to operate and how much excessive wear will it cause within the cylinder or core?

The truth is that many master key systems created by the original manufacturer may have problems. Many really bad systems have been sent out by factories when the wrong person achieved the position of Key Systems Manager, or worse yet, the position was left unfilled for too long. Additionally, duplicate systems have been sent out to the same customer for different buildings! The bad news is that as companies buy and sell other companies, it is not uncommon to fire employees whose knowledge is essential to the proper operation of the company, including Key Systems Managers, Master Keying Locksmiths, etc., within the "old" company. A new broom, contrary to the old adage, does not always "sweep clean".

LESSON EIGHT:
THE LIMITED PROGRESSION TECHNIQUE

(A subset of RCM that can be developed using TPP charts)

In the previous section, we saw that by pinning our cylinders to the Change Key combination and the bitting of the TMK, each 6 pin cylinder had 62 incidental keys that would also operate it IF THEY EXISTED. If it was a five pin cylinder, each would have 30 incidental master keys that would also operate it IF THEY EXISTED. If it had been a 7 pin cylinder, each would have been 126 incidental master keys per cylinder.

BUT, by assigning them to jobs as master keys of various levels, we were able to do away with the problems they pose. Right?

WRONG!!! This would only be true if the customer's system were so complex as to require each and every one of them. We have preserved security by NOT INCREASING the number of incidental keys, (as would happen if we stacked pins), but we have only DECREASED their number by any master keys that we ACTUALLY NEEDED for the system!!!

If, for example, the customer's system only required a Grand Master Key and 2 Master Keys under it, each cylinder would need to be operated by only its change key, the highest master key in the system, and ONE LOWER LEVEL MASTER KEY! So instead of having 62 phantom keys, we have 61. So what? That's not much of an increase in security!!!

So what is the solution? Are we locked into the loss of security? In most cases the answer is "NO"!!!

The answer lies in selecting APPROPRIATE pinning levels for the specific job at hand, including potential needs for anticipated or reasonably possible expansion.

If, for example, we needed only 4 change keys controlled by one master key, it would be folly indeed to use as our master the TMK!!! Instead, in a system where the KBA has 4 or 5 possibilities in the a position or c position, a BLOCK MASTER or a LINE MASTER (respectively) might be chosen to be the highest master key actually required by the system.

And if we're not going to use the higher master, why pin to it?

Instead, we pin only to the level of master key actually required by our system. For example, using the Sample chart we worked on earlier, we see that the Block Master for the first block on page one was 5 4 3 2 7 8, so our pinning would be:

BMK	**5 4 3 2 7 8**
CK	**5 4 3 2 7 0**
Smallest=Bottom Pin	**5 4 3 2 7 0**
Difference=Master Pin	*** * * * * 8**

Can you determine how many incidental master keys exist between the BMK and the Change Key in this cylinder?

If you determined the BMK to be the only master key of any level in that cylinder, you are correct.

That will not be the case every time this technique is applied, of course. The higher the key you choose to use as your Actual Top Master Key, the more Incidental Master Keys will remain.

For example, if we chose the VMK for our actual Top Master, that cylinder's pinning would be:

VMK	**5 4 3 2 5 8**
CK	**5 4 3 2 7 0**
Smallest=Bottom Pin	**5 4 3 2 5 0**
Difference=Master Pin	*** * * * * 2 8**

This would permit Incidental Master Keys (Don't ask why the abbreviation IMK has never caught on.) of 543250 and 543278 (the BMK) to operate in addition to the two keys we planned.

For example, if we chose the PMK for our actual Top Master, that cylinder's pinning would be:

PMK	**5 4 3 0 5 8**
CK	**5 4 3 2 7 0**
Smallest=Bottom Pin	**5 4 3 0 5 0**
Difference=Master Pin	*** * * 2 2 8**

This would permit Incidental Master Keys of 543250, 543050, 543258 (the VMK), 543078 (the HMK), 543070 (the Line Master) and 543278 (the BMK) to operate in addition to the two keys we planned.

Notice that any designated Incidental Master Keys below the actual chosen Top Master will remain in operation, but all Masters above it are removed from operational functionality. Now you see why this book referred to the TMK as the Theoretical Master Key.

When using this technique, the TMK may actually be taken out of the system and the system broken into many smaller systems instead.

REMEMBER: Pinning it to a higher level of master key automatically allows the lower levels of master key under it to also work their areas.

As you have just seen, we can pin the lock cylinders to any level of master key we need to. The fundamental rule is that we pin it to the HIGHEST LEVEL OF MASTER KEY ACTUALLY REQUIRED BY THE SYSTEM.

For now, let's look at its effect on security!!!

If we pin to a Block Master (4 or 5 Change Keys) as the highest master key in the system, how many incidental master keys exist for the lock cylinders involved? NONE!!! That's right, not a single unintended key will open any of those four locks. From 62 down to ZERO!!!

If we pin to a Vertical Column Master or a Horizontal Master (16 or 25 Change Keys) as the highest master key for the system, there are 2 unintended keys, and one of them is the Block Master. That's a BIG improvement.

If we pin the Page Master (64 or 125 Change Keys) as our highest master key in the system, each lock cylinder will have 6 unintended keys to operate it. From 62 to 6 is still excellent! Furthermore, one of those 6 is the Block Master, one is the Horizontal Master, one is the Horizontal Master and one is the Vertical Master.

Pinning to a Group Master (256 or 625 Change Keys) as the highest master key will allow 14 unintended keys to operate any given cylinder. From 62 to 14 is still a good increase in security, but one of the 14 is a Block Master, one is a Line Master, one is a Vertical Master, and one is a Horizontal Master, and one is the Page Master. This could conceivably decrease the number of unintended keys to as few as 9.

Pinning to the Section Master (1,024 or 3,125 Change Keys) doesn't give us as much an increase in security as the lower levels. It reduces it to 30, which is still about a 50% increase in security! Once again, one is the Block Master, one the Line Master, one the Vertical Master, one the Horizontal Master, one the Page Master, and one the Group Master (Base Master).

Whenever we pin every cylinder directly to a Change Key and any level of master key other than the theoretical master, we have created a system using the technique of Limited Progression!!!

However, it should not be used to "key into existing systems" unless you have enough knowledge of the system to be absolutely certain of its development and all the existing or prior change keys.

(The next lesson will explain why.)

LESSON NINE
THE ROTATING CONSTANT METHOD (RCM)

As we saw earlier, we can develop a system using Right to Left Progression or by varying the progression sequence. Varying it changed the order in which each combination appeared, and what logical groups they fell into for master keying purposes, but did not give us any new combinations.

This section deals with a slightly different METHOD of progressing the system, which typically gives us more combinations for change keys than does the TOTAL POSITION PROGRESSION or the LIMITED PROGRESSION TECHNIQUE.

The disadvantage of it is that we must be far more careful in breaking the system up into smaller master key groups, because each of the master keys involved has far greater power than in the systems built by the other two methods. Also, in the other methods, each master had relatively unique operational control of its change keys. In the method in this chapter, there is a certain amount of overlapping that we must watch out for in selecting our lower levels of master keys.

This method is known in the trade as **ROTATING CONSTANT** (Older terms no longer deemed appropriate are "**HOLD AND VARY**" and "**MOVABLE CONSTANT**.)

It should be mentioned, however, that methods NOT related to this one have also been known by the same exact names, to reflect an individual's teaching STYLE for the Total Position Progression. Care should thus be taken when discussing a system to determine the exact meaning of terms such as these, to avoid confusion.

The one restriction of the Hold and Vary method is that it can NEVER be mixed with a Total Position Progression Method master key system. The rules with which we develop a Rotating Constant system will AUTOMATICALLY cause the change keys to interchange drastically (as though they were masters) with the change keys from the other method.

However, this only applies to true TPP or Total Position Progression. It does not apply to the LIMITED PROGRESSION technique, even if it were originally written as a Total Position Progression system and then broken into smaller master key systems using Limited Progression.

Once you fully understand the Total Position Progression Method and the Limited Progression Method, it is easy to understand and develop a Rotating Constant system. The rules are only slightly different, and the math is virtually the same.

Our first step, as in the other methods, is to fill out the Possible Depths section of a KBA Chart. Then we select our increment size, select a TMK and using the chart and guidelines fill out the bottom (cut possibilities section of the chart.

In the Total Position Progression method, our next step was to select a progression sequence and begin development of the system. In the Hold and Vary Method, however, we will be breaking it up into several DIFFERENT progression sequences instead.

One of the rules of the Total Position Progression Method (but not the Limited Progression Method) was that the same depth cut never appeared in the same position on both the change key and the TMK. In the Rotating Constant Method, the exact OPPOSITE is true.

The "CONSTANT" refers to "holding constant" a CERTAIN NUMBER OF CUTS from the master key, and applying them to the change keys. Unlike the Limited Progression technique, however, it will not be the same cuts in the same positions on all the keys in the system.

BUT ALL THE CHANGE KEYS IN THE SYSTEM MUST HOLD THE SAME NUMBER OF CUTS FROM THE THEORETICAL MASTER!!!

That is an ABSOLUTE rule for Rotating Constant.

PROGRESSION PATTERNS FOR ROTATING CONSTANT SYSTEMS

As mentioned in the introduction to the Rotating Constant method, we will be "HOLDING CONSTANT" a certain number of cuts from the TMK on each key, although WHICH cuts were held will vary from key to key, as will which columns (or positions on the key) that we progress. In the Limited Progression Technique section we saw that the more positions from the TMK that were held, the fewer the number of possible change keys.

In order to understand this better, let's look at one group of progression patterns. Rotating Constant progression patterns are frequently identified by the numbers of the positions which we choose to hold. These positions are identified from bow to tip, as shown in this illustration:

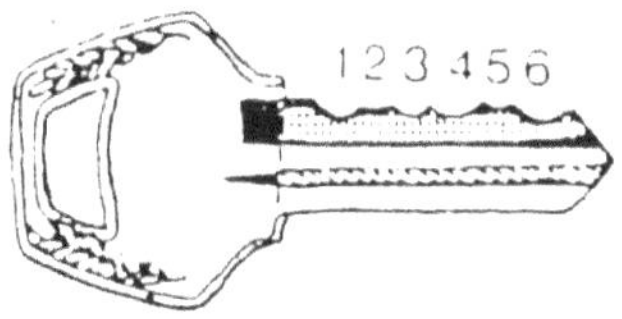

The first group we will look at consists of a six pin lock, where we have decided to hold five positions. For simplicity, we will assume that we have a manufacturer's numbering system of zero to nine to work with. Because we have six cuts on the key (for a six pin lock), and are "HOLDING CONSTANT" five positions, this means that we will be PROGRESSING one position. It's easy to see that we have six choices as to which position we will progress.

Let's assume a theoretical master of 238761. If we chose to progress only the sixth position, the base field would look like this:

2	3	8	7	6	1
					3
					5
					7
					9

and would give us the combinations:

2 3 8 7 6 3
2 3 8 7 6 5
2 3 8 7 6 7
2 3 8 7 6 9

with basic pinning of:

with basic pinning of:

```
MK 238761    MK 238761    MK 238761    MK 238761
CK 238763    CK 238765    CK 238767    CK 238769
BP 238761    BP 238761    BP 238761    BP 238761
MP -----2    MP -----4    MP -----6    MP -----8
```

If we chose to progress only the fifth position, the base field would look like this:

2	3	8	7	6	1
				0	
				2	
				4	
				8	

and would give us the combinations:

2 3 8 7 0 1
2 3 8 7 2 1
2 3 8 7 4 1
2 3 8 7 8 1

with basic pinning of:

```
MK 238761    MK 238761    MK 238761    MK 238761
CK 238701    CK 238721    CK 238741    CK 238781
BP 238701    BP 238721    BP 238741    BP 238761
MP ----6-    MP ----4-    MP ----2-    MP ----2-
```

If we chose to progress only the fourth position, the base field would look like this:

2	3	8	7	6	1
			1		
			3		
			5		
			9		

and would give us the combinations:

2 3 8 1 6 1
2 3 8 3 6 1
2 3 8 5 6 1
2 3 8 9 6 1

with basic pinning of:

MK 238761	MK 238761	MK 238761	MK 238761
CK 238161	CK 238361	CK 238561	CK 238961
BP 238161	BP 238361	BP 238561	BP 238761
MP ---6--	MP ---4--	MP ---2--	MP ---2--

If we chose to progress only the third position, the base field would look like this:

2	3	8	7	6	1
		0			
		2			
		4			
		6			

and would give us the combinations:

2 3 0 7 6 1
2 3 2 7 6 1
2 3 4 7 6 1
2 3 6 7 6 1

with basic pinning of:

MK 238761	MK 238761	MK 238761	MK 238761
CK 230761	CK 232761	CK 234761	CK 236761
BP 230761	BP 232761	BP 234761	BP 236761
MP --8---	MP --6---	MP --4---	MP --2---

If we chose to progress only the second position, the base field would look like this:

2	3	8	7	6	1
	1				
	5				
	7				
	9				

and would give us the combinations:

2 1 8 7 6 1
2 5 8 7 6 1
2 7 8 7 6 1
2 9 8 7 6 1

with basic pinning of:

MK 238761	MK 238761	MK 238761	MK 238761
CK 218761	CK 258761	CK 278761	CK 298761
BP 218761	BP 238761	BP 238761	BP 238761
MP -2----	MP -2----	MP -4----	MP -6----

If we chose to progress only the first position, the base field would look like this:

2	3	8	7	6	1
0					
4					
6					
8					

and would give us the combinations:

0 3 8 7 6 1
4 3 8 7 6 1
6 3 8 7 6 1
8 3 8 7 6 1

with basic pinning of:

MK 238761	MK 238761	MK 238761	MK 238761
CK 038761	CK 438761	CK 639761	CK 838761
BP 038761	BP 238761	BP 238761	BP 238761
MP 2-----	MP 2-----	MP 4-----	MP 6-----

Notice that in each of those six possible systems that we developed, the theoretical master controlled 4 combinations, each of which, when the lock was pinned, would only have one master pin in it.

BUT WAIT A MINUTE! The master didn't change a bit, and none of the change keys we developed would interchange with any of the others, even though the master would operate all of them!

Congratulations! You've just seen a Rotating Constant system developed from scratch. This would be a six pin system HOLDING FIVE POSITIONS CONSTANT from the TMK. There are actually six progression patterns involved:

12345 12346 12356 12456 13456 23456

(The patterns are identified by the positions where we "HOLD" a master cut.)
which could also be written as:

CCCCC**P** CCCC**P**C CCC**P**CC CC**P**CCC C**P**CCCC **P**CCCCC

Where C indicates a position we have held the master cut CONSTANT for, and **P** indicates a position where we will PROGRESS the cuts, the following is true:

Pattern 12345 would be CCCCC**P**
Pattern 12346 would be CCCC**P**C
Pattern 12356 would be CCC**P**CC
Pattern 12456 would be CC**P**CCC
Pattern 13456 would be C**P**CCCC
Pattern 23456 would be **P**CCCCC

This method gave us 24 combinations for change keys under one master, with only one master pin per lock! In the Total Position Progression method or using the Limited Progression Technique, on the same brand of locks, with the same master key pattern, one master pin per lock would have given us only 4 possible change keys. Two master pins per lock would have given us only 16 possible change keys. In fact, to control 24 possible change keys, we would have had to put THREE master pins in each and every lock in the system! As we saw earlier in the book, the more master pins in the lock, the less security the lock has. The less master pins, the better security the lock has. So the Rotating Constant Method, on a simple system like this, increases both the number of possible change keys, and the security of the lock!

The next two pages show the progression patterns for a six pin lock cylinder and for a five pin lock cylinder.

The number of change keys that we end up with under Rotating Constant will depend on the numbering system the lock manufacturer uses, whether the locks are 5 pin or 6 pin, how many positions we "HOLD CONSTANT" from the TMK, and the options we chose earlier.

REMEMBER THAT WE CANNOT COMBINE PATTERNS THAT HOLD A DIFFERENT NUMBER OF POSITIONS, INTO THE SAME SYSTEM. WE MUST EITHER "HOLD ONE" OR "HOLD TWO" OR ETC. (There are two variants on the RCM technique that do not break this rule but at first glance may seem to. We will cover them in the following lesson.)

The progression patterns for a six pin lock would be:

HOLDING ONE (POSITION) (5 master pins per lock)

1 C**PPPPP**
2 **P**C**PPPP**
3 **PP**C**PPP**
4 **PPP**C**PP**
5 **PPPP**C**P**
6 **PPPPP**C

HOLDING TWO (POSITIONS) (4 master pins per lock)

12 CC**PPPP**
13 C**P**C**PPP** 23 **P**CC**PPP**
14 C**PP**C**PP** 24 **P**C**P**C**PP** 34 **PP**CC**PP**
15 C**PPP**C**P** 25 **P**C**PP**C**P** 35 **PP**C**P**C**P** 45 **PPP**CC**P**
16 C**PPPP**C 26 **P**C**PPP**C 36 **PP**C**PP**C 46 **PPP**C**P**C 56 **PPPP**CC

HOLDING THREE (POSITIONS) (3 master pins per lock)

123 CCC**PPP**
124 CC**P**C**PP**
125 CC**PP**C**P**
126 CC**PPP**C
134 C**P**CC**PP** 234 **P**CCC**PP**
135 C**P**C**P**C**P** 235 **P**CC**P**C**P**
136 C**P**C**PP**C 236 **P**CC**PP**C
145 C**PP**CC**P** 245 **P**C**P**CC**P** 345 **PP**CCC**P**
146 C**PP**C**P**C 246 **P**C**P**C**P**C 346 **PP**CC**P**C
156 C**PPP**CC 256 **P**C**PP**CC 356 **PP**C**P**CC 456 **PPP**CCC

HOLDING FOUR (POSITIONS) (2 master pins per lock)

1234 CCCC**PP**
1235 CCC**P**C**P**
1236 CCC**PP**C
1245 CC**P**CC**P**
1246 CC**P**C**P**C
1256 CC**PP**CC
1345 C**P**CCC**P** 2345 **P**CCCC**P**
1346 C**P**CC**P**C 2346 **P**CCC**P**C
1356 C**P**C**P**CC 2356 **P**CC**P**CC
1456 C**PP**CCC 2456 **P**C**P**CCC 3456 **PP**CCCC

HOLDING FIVE (POSITIONS) (1 master pin per lock)

12345 CCCCC**P** 12456 CC**P**CCC
12346 CCCC**P**C 13456 C**P**CCCC
12356 CCC**P**CC 23456 **P**CCCCC

The progression patterns for a five pin lock would be:

HOLDING ONE (POSITION) (4 master pins per lock)
1 C**PPPP**
2 **P**C**PPP**
3 **PP**C**PP**
4 **PPP**C**P**
5 **PPPP**C

HOLDING TWO (POSITIONS) (3 master pins per lock)
12 CC**PPP**
13 C**P**C**PP** 23 **P**CC**PP**
14 C**PP**C**P** 24 **P**C**P**C**P** 34 **PP**CC**P**
15 C**PPP**C 25 **P**C**PP**C 35 **PP**C**P**C 45 **PPP**CC

HOLDING THREE (POSITIONS) (2 master pins per lock)
123 CCC**PP**
124 CC**P**C**P**
125 CC**PP**C
134 C**P**CC**P** 234 **P**CCC**P**
135 C**P**C**P**C 235 **P**CC**P**C
145 C**PP**CC 245 **P**C**P**CC 345 **PP**CCC

HOLDING FOUR (POSITIONS) (1 master pin per lock)
1234 CCCC**P**
1235 CCC**P**C
1245 CC**P**CC
1345 C**P**CCC 2345 **P**CCCC

Now let's look at a sample system worked out for Holding 4 positions Constant of 6. We can use the standard charts as described on the previous page.

Note that each progressed position is progressed, not in terms of its position on the key, but in terms of its progression sequence in that particular Rotating Constant Pattern.

Bear in mind that in this pattern, there are not any Horizontal Masters, Line Masters, or Page Masters. The highest key in this system is the TMK. Each of the keys in the system has only two cuts different from it. Therefore, it could be said in this particular example, that the TMK is a Vertical Master, although as we see, it operates multiple Vertical columns (one per pattern on pattern) instead of the more traditional one column from TPP or using the Limited Progression technique.

Therefore, there can be no additional Level 2 masters. The Block Master capability is still present, though, as you will see shortly, but with changed capabilities.

The progression would have been the same if we had used standard charts, as long as we keep in mind which progression pattern we are dealing with. It usually helps to write it above the particular chart you are using. Remember that you get a new First Progression Sequence, Second Progression Sequence, etc with each chart. They will not be in the same positions from chart to chart.

Remember that if the pattern is "Progressing 1", each progression pattern will require only a Block to progress. If it is "Progressing 2", as in this example, it will require a Vertical Column for each pattern. If it is "Progressing 3", it will require a complete page for each progression pattern, etc.

Let's look at a RCM Pattern of Holding 4 Constant and Progressing 2 on a six pin (also known as Holding 4 of 6) and see how the 15 patterns develop.

Remember that although we are using a standard SPF chart to write them, we are actually writing 4 separate charts (progression patterns) on each sheet.

5 4 3 2 5 8

VERTICAL MASTER

C	C	C	C	P	P
5	4	3	2	7	0
5	4	3	2	7	2
5	4	3	2	7	4
5	4	3	2	7	6

BLOCK MASTER

C	C	C	P	C	P
5	4	3	4	5	0
5	4	3	4	5	2
5	4	3	4	5	4
5	4	3	4	5	6

BLOCK MASTER

C	C	P	C	C	P
5	4	5	2	5	0
5	4	5	2	5	2
5	4	5	2	5	4
5	4	5	2	5	6

BLOCK MASTER

C	P	C	C	C	P
5	6	3	2	5	0
5	6	3	2	5	2
5	6	3	2	5	4
5	6	3	2	5	6

BLOCK MASTER

5	4	3	2	9	0
5	4	3	2	9	2
5	4	3	2	9	4
5	4	3	2	9	6

BLOCK MASTER

5	4	3	6	5	0
5	4	3	6	5	2
5	4	3	6	5	4
5	4	3	6	5	6

BLOCK MASTER

5	4	7	2	5	0
5	4	7	2	5	2
5	4	7	2	5	4
5	4	7	2	5	6

BLOCK MASTER

5	8	3	2	5	0
5	8	3	2	5	2
5	8	3	2	5	4
5	8	3	2	5	6

BLOCK MASTER

5	4	3	2	1	0
5	4	3	2	1	2
5	4	3	2	1	4
5	4	3	2	1	6

BLOCK MASTER

5	4	3	8	5	0
5	4	3	8	5	2
5	4	3	8	5	4
5	4	3	8	5	6

BLOCK MASTER

5	4	9	2	5	0
5	4	9	2	5	2
5	4	9	2	5	4
5	4	9	2	5	6

BLOCK MASTER

5	0	3	2	5	0
5	0	3	2	5	2
5	0	3	2	5	4
5	0	3	2	5	6

BLOCK MASTER

5	4	3	2	3	0
5	4	3	2	3	2
5	4	3	2	3	4
5	4	3	2	3	6

BLOCK MASTER

5	4	3	0	5	0
5	4	3	0	5	2
5	4	3	0	5	4
5	4	3	0	5	6

BLOCK MASTER

5	4	1	2	5	0
5	4	1	2	5	2
5	4	1	2	5	4
5	4	1	2	5	6

BLOCK MASTER

5	2	3	2	5	0
5	2	3	2	5	2
5	2	3	2	5	4
5	2	3	2	5	6

BLOCK MASTER

P	C	C	C	C	P
7	4	3	2	5	0
7	4	3	2	5	2
7	4	3	2	5	4
7	4	3	2	5	6

BLOCK MASTER

C	C	C	P	P	C
5	4	3	4	7	8
5	4	3	4	9	8
5	4	3	4	1	8
5	4	3	4	3	8

BLOCK MASTER

C	C	P	C	P	C
5	4	5	2	7	8
5	4	5	2	9	8
5	4	5	2	1	8
5	4	5	2	3	8

BLOCK MASTER

C	P	C	C	P	C
5	6	3	2	7	8
5	6	3	2	9	8
5	6	3	2	1	8
5	6	3	2	3	8

BLOCK MASTER

9	4	3	2	5	0
9	4	3	2	5	2
9	4	3	2	5	4
9	4	3	2	5	6

BLOCK MASTER

5	4	3	6	7	8
5	4	3	6	9	8
5	4	3	6	1	8
5	4	3	6	3	8

BLOCK MASTER

5	4	7	2	7	8
5	4	7	2	9	8
5	4	7	2	1	8
5	4	7	2	3	8

BLOCK MASTER

5	8	3	2	7	8
5	8	3	2	9	8
5	8	3	2	1	8
5	8	3	2	3	8

BLOCK MASTER

1	4	3	2	5	0
1	4	3	2	5	2
1	4	3	2	5	4
1	4	3	2	5	6

BLOCK MASTER

5	4	3	8	7	8
5	4	3	8	9	8
5	4	3	8	1	8
5	4	3	8	3	8

BLOCK MASTER

5	4	9	2	7	8
5	4	9	2	9	8
5	4	9	2	1	8
5	4	9	2	3	8

BLOCK MASTER

5	0	3	2	7	8
5	0	3	2	9	8
5	0	3	2	1	8
5	0	3	2	3	8

BLOCK MASTER

3	4	3	2	5	0
3	4	3	2	5	2
3	4	3	2	5	4
3	4	3	2	5	6

BLOCK MASTER

5	4	3	0	7	8
5	4	3	0	9	8
5	4	3	0	1	8
5	4	3	0	3	8

BLOCK MASTER

5	4	1	2	7	8
5	4	1	2	9	8
5	4	1	2	1	8
5	4	1	2	3	8

BLOCK MASTER

5	2	3	2	7	8
5	2	3	2	9	8
5	2	3	2	1	8
5	2	3	2	3	8

BLOCK MASTER

P	C	C	C	P	C
7	4	3	2	7	8
7	4	3	2	9	8
7	4	3	2	1	8
7	4	3	2	3	8

BLOCK MASTER

C	C	P	P	C	C
5	4	5	4	5	8
5	4	5	6	5	8
5	4	5	8	5	8
5	4	5	0	5	8

BLOCK MASTER

C	P	C	P	C	C
5	6	3	4	5	8
5	6	3	6	5	8
5	6	3	8	5	8
5	6	3	0	5	8

BLOCK MASTER

P	C	C	P	C	C
7	4	3	4	5	8
7	4	3	6	5	8
7	4	3	8	5	8
7	4	3	0	5	8

BLOCK MASTER

9	4	3	2	7	8
9	4	3	2	9	8
9	4	3	2	1	8
9	4	3	2	3	8

BLOCK MASTER

5	4	7	4	5	8
5	4	7	6	5	8
5	4	7	8	5	8
5	4	7	0	5	8

BLOCK MASTER

5	8	3	4	5	8
5	8	3	6	5	8
5	8	3	8	5	8
5	8	3	0	5	8

BLOCK MASTER

9	4	3	4	5	8
9	4	3	6	5	8
9	4	3	8	5	8
9	4	3	0	5	8

BLOCK MASTER

1	4	3	2	7	8
1	4	3	2	9	8
1	4	3	2	1	8
1	4	3	2	3	8

BLOCK MASTER

5	4	9	4	5	8
5	4	9	6	5	8
5	4	9	8	5	8
5	4	9	0	5	8

BLOCK MASTER

5	0	3	4	5	8
5	0	3	6	5	8
5	0	3	8	5	8
5	0	3	0	5	8

BLOCK MASTER

1	4	3	4	5	8
1	4	3	6	5	8
1	4	3	8	5	8
1	4	3	0	5	8

BLOCK MASTER

3	4	3	2	7	8
3	4	3	2	9	8
3	4	3	2	1	8
3	4	3	2	3	8

BLOCK MASTER

5	4	1	4	5	8
5	4	1	6	5	8
5	4	1	8	5	8
5	4	1	0	5	8

BLOCK MASTER

5	2	3	4	5	8
5	2	3	6	5	8
5	2	3	8	5	8
5	2	3	0	5	8

BLOCK MASTER

3	4	3	4	5	8
3	4	3	6	5	8
3	4	3	8	5	8
3	4	3	0	5	8

BLOCK MASTER

LESSON TEN: Adding Master Keys for Rotating Constant Systems

The first thing to understand in breaking a Rotating Constant system into smaller master key groups is that the number of positions we are progressing will be the first limitation we face.

As mentioned in the previous lesson, **when Progressing 1**, the progression can be accomplished for each pattern on a Block. Therefore, the Theoretical Master is in reality a Block Master, and there are no submasters of any level except that Block Master.

When Progressing 2, the progression can be accomplished using a Vertical Column for each progression pattern, and there are 15 patterns for a six pin key. Therefore, the Theoretical Master is in reality a Vertical Master. The only submasters are the Block Masters. There are additional incidental master keys that operate the cylinder, but they are not clearly defined masters such as the block master. Each pattern yields 16 Change Keys, (25 for a single increment with 5 possibilities in the KBA) as it normally would under the Vertical Master.

When Progressing 3, the progression can be accomplished using a full page for each progression pattern, and there are 20 patterns for a six pin key. Therefore, the Theoretical Master is in reality a Page Master, and we have the capability of Block Masters, Vertical Masters, Horizontal Masters, and Line Masters under it, as well as a few undefined incidental master keys. Each Pattern yields 64 Change Keys (125 for a single increment with 5 possibilities in the KBA)

When Progressing 4, we will need a Group of pages to accomplish it. Each Pattern yields 256 Change Keys (625 for a single increment with 5 possibilities in the KBA)

Progressing 5 requires a Section of pages to accomplish each progression pattern. Each Pattern yields 1,024 Change Keys (3,125 for a single increment with 5 possibilities in the KBA)

Each Master must be filled in on the chart before any selection of appropriate masters is made, since it will then be discovered that what was a Block Master in one progression pattern of Progressing 3, for example, will also be a Line Master in another progression, another Block Master in another, and especially difficult to spot, may operate as a sort of Vertical Line Master in another.

This is because a Block Master in Holding 4 Progressing 2, for example, will end up having 5 out of six cuts in common with the Theoretical Master. It will be limited only by one cut position, which will give it its assigned level of authority within the system.

If it is a Block Master for the first block in the first column, everything else it is will also be a "first", e.g. the first line or the first combination in the each block in the first column.

In a progression pattern Progressing 3, the Block and Line Masters will follow that same Pattern, the Vertical and Horizontal Masters will typically still operate one column or none per chart, but may also have a different assignment – one combination in each of the sixteen blocks on a page. Once again, if they are the first column in one chart, for a chart on which they operate one bitting per block it will be the first combination. If they are the second column (horizontal or vertical) on one chart and there is another chart on which they operate one bitting per block, it will be the second bitting. They NEVER jump positions.

If you look through the chart, you will see that several of the masters are the same, even though the combinations in the Block, Column, etc. that they control is NOT the same.

These are the first keys you can add to the list of those operated by that master key. You will see a few things about the Masters if you look:

1) SIZE: Although there will be some intertwining of operational control between Line Masters and Block Masters, and some between Vertical and Horizontal Masters, the keys will never "JUMP" a level. The bitting for a Block Master, for example, will never be the same as that for a Vertical or Horizontal Master.

2) POSITION: When a Block Master operates a full block, the position of that block will determine the position of all the other keys it operates. For example, if it operates the full block in the first position on a page, all the other keys it operates will be "firsts" also: first block on other pages, first line across the top of a page, or first combination in top of a block in each of the blocks in a Vertical Column.

If it controlled the second Block on a page, it would control the second block on other pages, second line across the top of a page or pages, and second combination in each block in a Vertical Column.

Similarly, if a Vertical Master operates the first column on a page, it will also control the first column on other pages, will be the first Horizontal Master on other pages, and will control the first line across the top of the Blocks on each page.

How do we determine which combinations are controlled by a particular Master? Of course, it's obvious when it is one of the control groups where we have the master calculated. If the master has the same bitting, it will control the group.

But what about the groups such as the "second combination in each block of a Vertical Column"? How do we spot these?

It is actually very easy. When looking for the control groups of the various Block Masters or Line Masters in a "Progressing 2" progression pattern, each Block Master will operate every key that has the same bitting as it has in its SECOND PROGRESSION SEQUENCE.

When looking for the control group of a Block Master or Line Master in a "Progressing 3" progression pattern, it will control all the keys that have the same cut as it has in the SECOND & THIRD PROGRESSION SEQUENCE.

When looking for the control groups of the various Horizontal and Vertical Masters in a "Progressing 3" progression pattern, it will control all the keys that have the same cut as it has in its third progression sequence.

When looking for the control groups of the various Vertical or Horizontal Masters in a "Progressing 4" progression pattern, it will control all the keys that have the same cut it has in the 3rd and 4th PROGRESSION SEQUENCES.

When looking for the control groups of the various Page Masters in a "Progressing 4" progression pattern, it will control all the keys that have the same cut it has in its FOURTH PROGRESSION SEQUENCE.

When looking for the control groups of the various Block or Line Masters in a "Progressing 4" progression pattern, it will control all the combinations that have the same cut it has in its 2nd, 3rd, & 4th PROGRESSION SEQUENCES.

When looking for the control groups of a Base Master or Group Master, in a "Progressing 5" progression pattern, it will control all the keys with the same cut it has in its 5th PROGRESSION SEQUENCE.

In a "Progressing 5" progression pattern, the Page master operates all with the same cut in the 4th & 5th SEQUENCES. The Vertical or Horizontal Master operates all the combinations with the same cut it has in its 3rd, 4th, & 5th SEQUENCES. The Block or Line Master operates all with the same cut it has in its 2nd, 3rd, 4th, & 5th PROGRESSION SEQUENCES.

For all of these, look for the cut that LIMITS its operation. That is the cut that defines its operational authority.

This is an ADVANCED technique and should not be attempted without an advanced skill set in master keying.

There are books specifically on the topic of the *Rotating Constant Method*, as well as *The Security Professional's Guide to Master Keying*, which goes into great depth on the topic.

The first book to mention it was *The Master Plan*, which was the first master keying book offered by Billy B. Edwards. It was then incorporated into his *Master Keying by the Numbers*, which has been updated about every fifteen years since, and is now in its fourth edition (2007).

It is recommended that you study the topic in depth before starting any new systems utilizing it.

LESSON ELEVEN:
The XCM Variant of the RCM

Another way exists to assign a limited number of master keys that do not overlap each other. It was developed by Jerome Andrews, modified slightly by A.J. Hoffman and practiced by Schlage lock for several years. However, it was shown as a "trick" and as such did not draw the attention it deserves. This was pointed out to this author in 2005 by Lloyd Seliber, a former Schlage training department employee and a former associate of A.J. Hoffman. In truth, it is an amazingly effective tool, and yet quite simple.

The technique begins by splitting the key into "Master" positions and "Change Key" positions. This has also been done as a subset of the Rotating Constant Method, whereby only a limited portion of the key is progressed (similar to an expanded version of the Limited Progression method.)

One might, for example, choose to progress "One in Three", where only three chambers are ever progressed, and only one position is progressed in each pattern, or "Two in Three of Six" whereby once again only three of the six possible positions are ever progressed, but in this case it two chambers at a time are progressed.

Indeed, it can be argued that one example of the Limited Progression we saw earlier is simply a "Three of Three in Six" RCM (Rotating Constant Method) system, where three positions are progressed but only three are ever progressed.

In the XMC variant, for a six pin key, this might typically be three positions for Change Keys and two for Master Keys. The sixth position is Held Constant. Once the system is in place, it can be decided, for future expansion needs, whether to consider that position one of the Master positions, and progress those positions as a Holding 1 Constant out of three on the six pin key, or to expand the number of Change Keys by rotating it as a Holding one out of four on a six pin key with the Change Key positions.

If you have a seven pin key, you have even more possibilities and could even use more than one position as the X position to be held constant and then associated with either Masters (M) or Change Keys (C).

On a six pin you are starting out with 16 (or 25 with a five cut KBA such as Single Increment) Master Keys under the TMK, which can be expanded by applying Rotating Constant to the Master positions to three times that many, or 54 (75) Master Key Levels.

On a six pin key you are starting out with 64 (125) Change Keys per Master Key, with the capability to expand that to 256 (625) by applying Rotating Constant to the Change Key positions to four times that many.

This is an ADVANCED technique and should only be attempted by those thoroughly understanding the technique. The book ***Advanced Master Keying Skills*** by Don O'Shall, Tony West, John Truempy, and Greg Brockway, **ISBN # 9774511-2-7** is an excellent resource for this particular topic.

The 2008 book, *The Security Professional's Guide to Master Keying*, by Don O'Shall, Tony West, John Truempy and Greg Brockway is perhaps the **most in depth treatise on the topic**, and well worth the price for the professional Master Keying Locksmith or Institutional Locksmith. This book is a complete overhaul from its original title (The Complete Security Professional's Guide to Master Keying) which was removed from the market by its authors shortly after its release because Master Keying professionals felt the Locksmithing public was not ready for such a book yet. The removal of the word "complete" occurred as the authors realized that Master Keying is continuing to evolve as more specialists research the topic, and that it will likely always be a work in progress and never truly complete.

LESSON TWELVE:
MULTIPLEX KEY SYSTEMS

Up until now, this book has dealt with the parameters of a professional master key system on one keyway. Certain manufacturers, however, utilize a "**Multiplex KEY SYSTEM**" to allow for expansion beyond the normal limits of the system.

A multiplex key system is a group of keyways designed by the manufacturer so that, while each keyway is designed not to interchange with any of the others in the system, there are enough common factors in the design of them, that certain key blanks, designed to utilize these common factors, can pass some or all of these keyways. In other words, there are "master blanks" that, if properly cut, would operate ALL the locks in the system, in spite of the different keyways involved.

If there were nine individual keyways, (not including master blank keyways, since the master keyways are only cut on the blank, not into the plugs) this would expand the possible combinations by nine times. On a lock with ten possible depths,(0 to 9 or 1 to 0) we had 4096 POSSIBLE change keys using a six pin key. Using a multiplex keyway system with nine keyways, we would instead have 36,864 possible change keys! Impressive. right?

Now for the bad news.

The security of these systems lies in ONLY the correct keys being cut on the master blanks. However, key cutters in five and dime stores, discount department stores, auto parts stores, (just to name a few -- I have even found this happening in Pizza Shoppes, and one Bridal Gown store) do NOT STOCK ALL THE KEYS IN THE GROUP!!! Instead they stock ONLY the master key blank that passes ALL the keyways in the group! No matter WHAT keyway the customer's key comes in, it is duplicated onto the "GRAND MASTER KEY BLANK"!!!

This destroys the security of the whole system! I have even seen a few locksmiths do this!

IN AREAS WHERE LOCKSMITHS AND KEY MAKERS ARE LICENSED, THIS IS A SEVERE CRIME!!! Unfortunately, in most areas, UNLESS THE KEY CAN BE TRACED to a particular locksmith or keymaker AFTER it has been used for a crime, it is usually only UNETHICAL!!! The fact that it is, at the least, unethical, seems to have had no effect on the number of times it occurs!

Some of the multiplex key groups, however, have not become popular enough for these keycutters to bother stocking them, thus, at least temporarily ensuring their security.

The following illustrations will show some of the multiplex key systems currently in use and the way the blanks are designed. Remember that only the top row is an actual keyway for the lock cylinder plugs.

The "MASTER" keyways are TYPICALLY KEY BLANKS ONLY--not lock cylinders! (exceptions exist).

If you look closely at any of the keyways for the primary keys (ones the core or cylinder is actually broached to accept properly, not the master blanks' keyways) you can usually see the pattern of how the keys work.

As an example, on the Corbin 59 series, there are two grooves on one side and on groove on the other. These grooves either go "up" or "down" while the master keyway is cut round to bypass both.

The following will show these keyways and the method of determining which keyway you have for a core or cylinder or key.

The basics of this keyway multiplex system are:

If the top groove on the side with two grooves goes "up" it is either C or D, and if it goes down it is either A or B.

If the bottom groove on that side goes up it is a 2, and if it goes down it is a 1.

The single groove in the middle of the key on the other side differentiates between C and D or A and B. If it goes up it is a B or D, and if it goes down it is an A or C.

Available 5, 6 and 7-pin
Z Class bitting (pages 62, 63 and 64)
60 is the reverse of 70

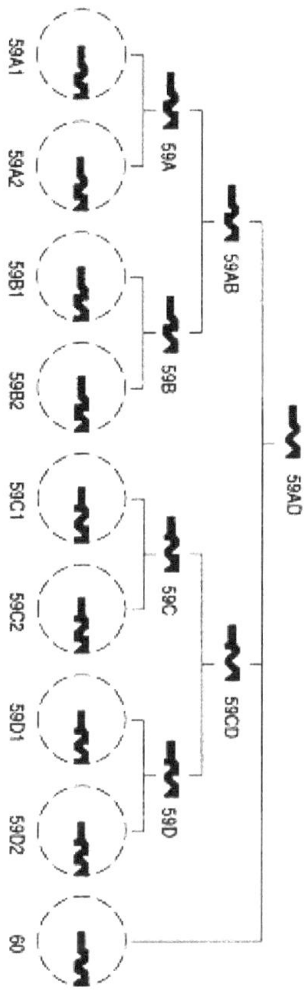

(Not shown actual size)

So the Possible Patterns are:

Left upper	Left Lower	Right Center	Keyway
D	D	D	59A1
D	U	D	59A2
D	D	U	59B1
D	U	U	59B2
U	D	D	59C1
U	U	D	59C2
U	D	U	59D1
U	U	U	59D2
V	V	V	60

Master Key Blank Keyway Possible Patterns Are:

D	R	D	O59A
D	R	U	O59B
U	R	D	O59C
U	R	U	059D
D	R	R	59AB
U	R	R	59CD
R	R	R	59AD

Therefore, if we have a key where the two on one side both go down that indicates either A1 or B1, and if the groove in the middle on the other side goes up it determines that it is a 59B1 keyway.

By the way the top master keyway, 059AD also passes the 60 keyway, which uses a V cut in each of the three positions, which is passed by the round cut in all three positions. It does NOT pass the 70 keyway, because it is the REVERSE of the 60, and therefore has the cut positions on the wrong side of the key.

On Sargent double lettered series primary keyways, LA-LC, LE-LG, and LL-LL, RA-RL, etc., if the bottom groove is "down" and on the left side of the blank (thumb side if holding the blank in your right hand), the keyway is LA, LB or LC. If the bottom groove is "up" and on the left, it is LE, LF or LG and if it is a V shape, it is LJ, LK, or LL.

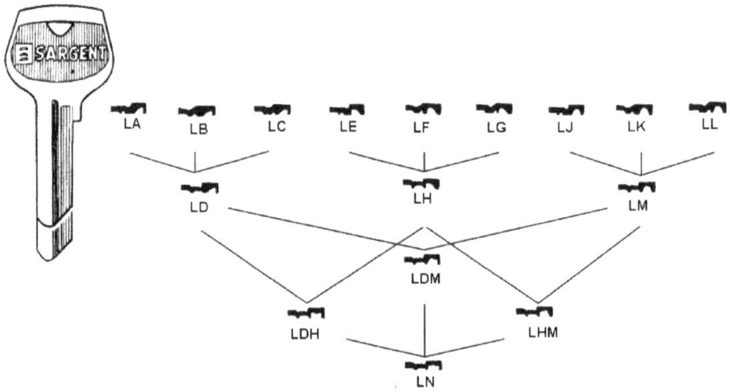

If on the other hand it is on the right and "down" the second letter possibilities remain the same but it is an RA, etc.

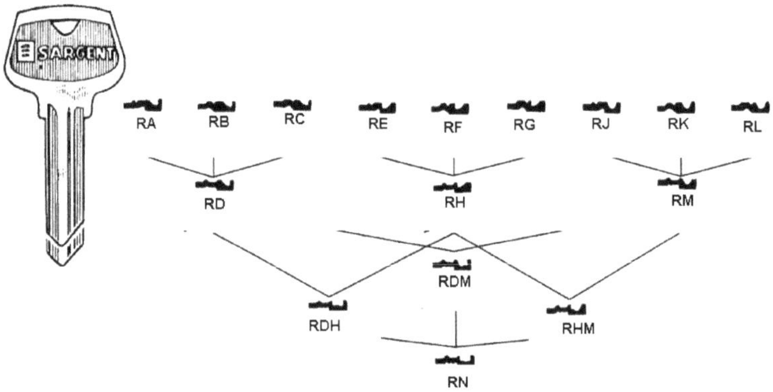

The three positions at the top portion of the keyway determine the exact letter between the three possibilities. There are three places for a small square groove to be placed: Top, Middle and Bottom. Top position indicates A, E, or J. The Middle position indicates B, F, or K. The Lower position indicates C, G, or L.

Understand that the "master keyways" are key blanks only, NOT cylinder keyways. Cylinders are typically only available broached in the primary keyways. Schlage is an exception to this rule. They have an H keyway available that the key passes the C, CE and E keyways, for example.

The following illustrations will show some of the multiplex key systems currently in use and the way the blanks are designed. Remember that only the top row is an actual keyway for the lock cylinder plugs.

The "MASTER" keyways are typically KEY BLANKS ONLY--not lock cylinder keyways!

But there are exceptions to that guideline.

For example Schlage Classic Keyways offered the C, E and H keyways for both plugs and keys, but the H keyway keys passed the C and E, as shown in the following illustration.

Schlage Classic Keyway Multiplex System

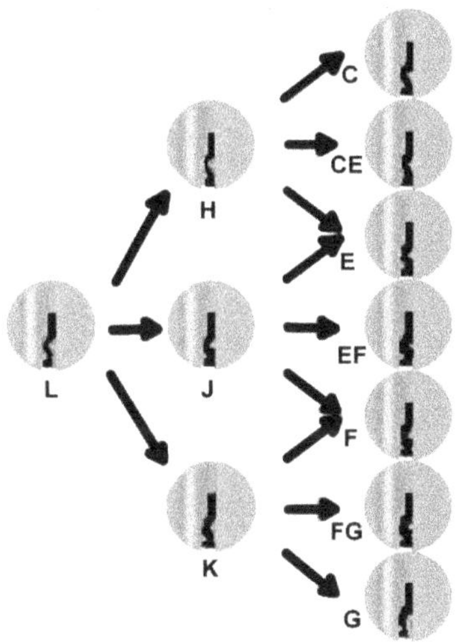

(not shown actual size)

Of the above keyways, only the L keyway is not available as a cylinder keyway. This manufacturer's Classic Keyway line is one of the few exceptions to the rule that the "MASTER" keyways are typically KEY BLANKS ONLY--not lock cylinder keyways!

By the way, the list glossary refers to **multiplex keyway systems** as "**multiplex**", and systems not capable of this as "**simplex**". At the time of this writing the term SIMPLEX has yet to be in common usage. As said throughout the book, learn and use all the applicable terms. It will benefit you.

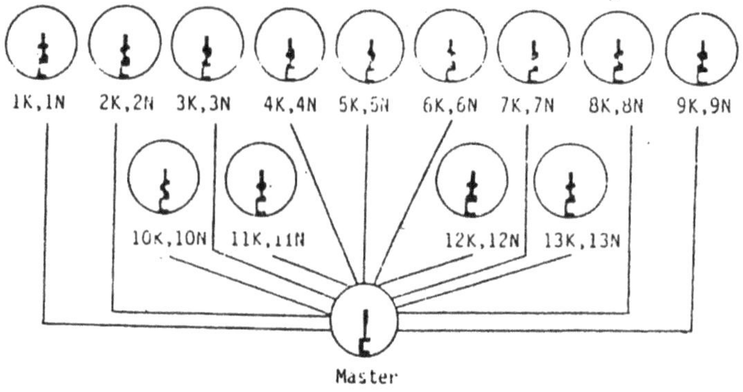

RUSSWIN
980 SERIES
960 SERIES

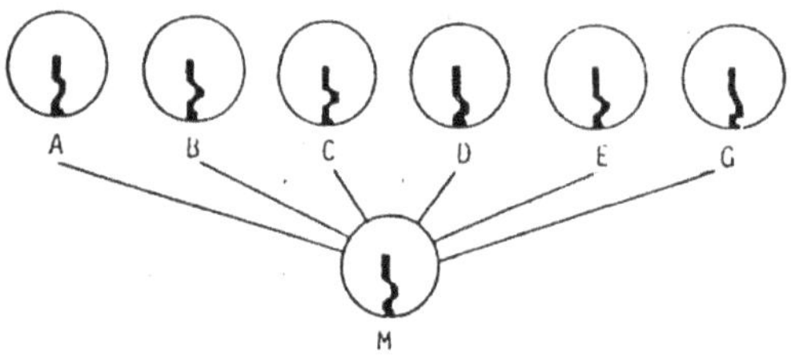

BEST A-M
FALCON
IC

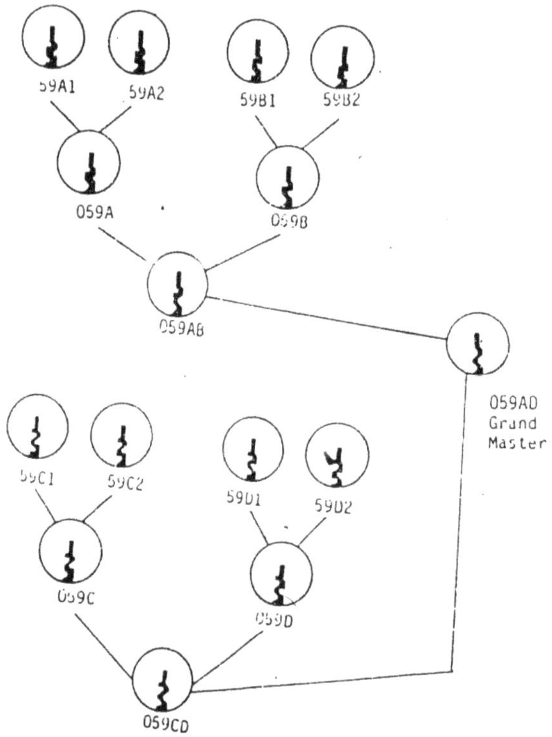

CORBIN 59 SERIES

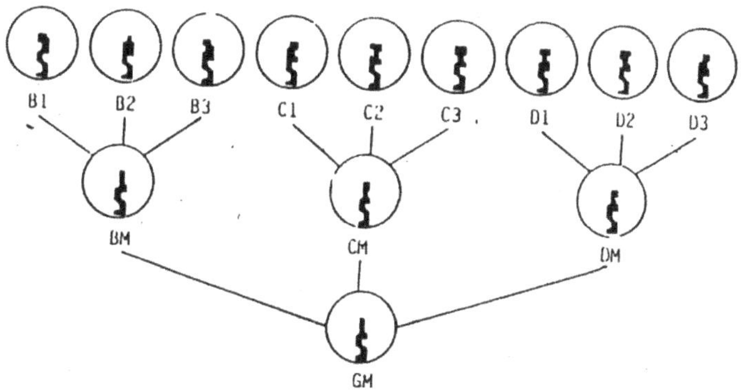

DEXTER
67 SERIES (5 PIN)
62 SERIES (6 PIN)

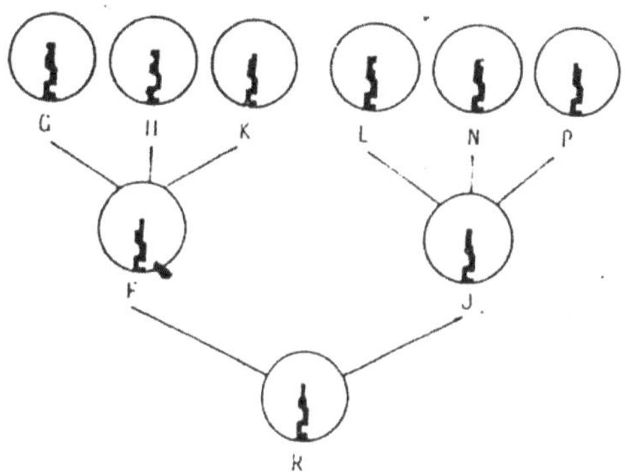

Falcon/Weiser (REGULAR)

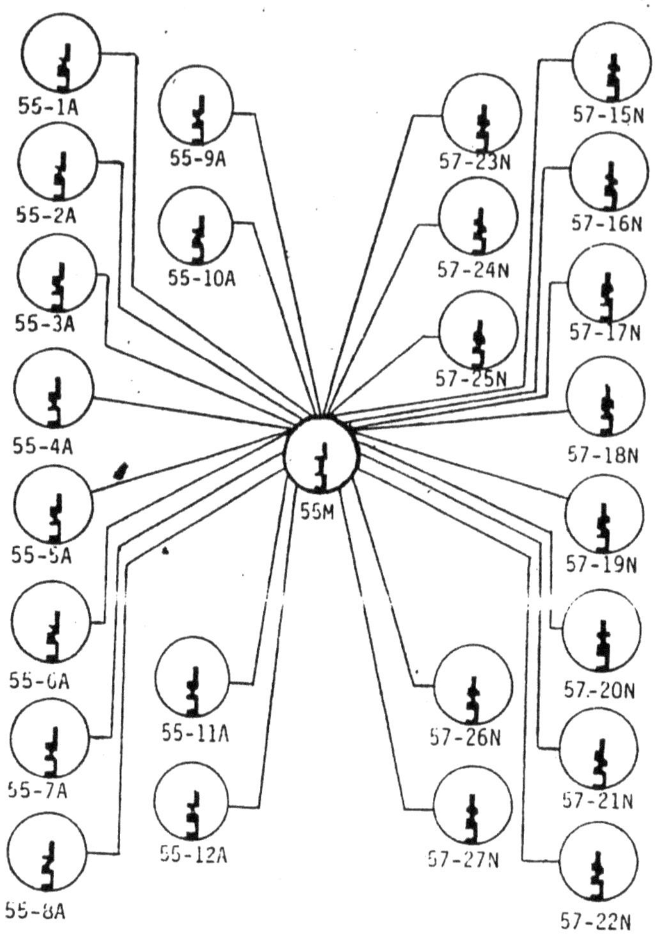

RUSSWIN
752 SERIES
6960 SERIES
(Taylor #'s Shown)

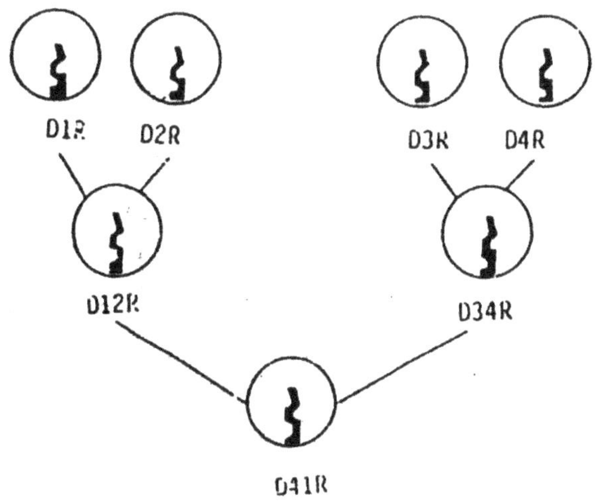

RUSSWIN

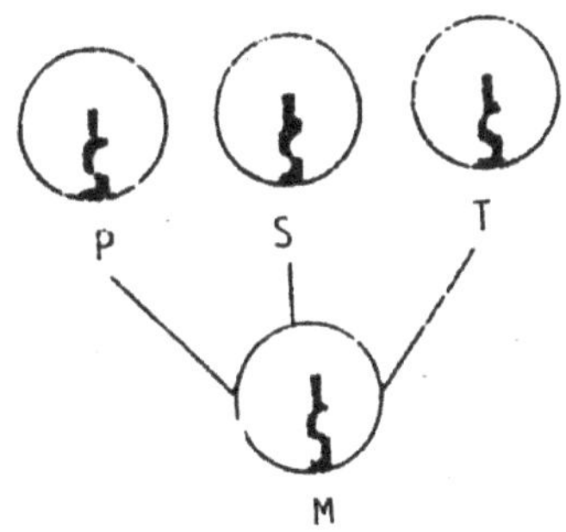

RUSSWIN

Sargent LA-LN

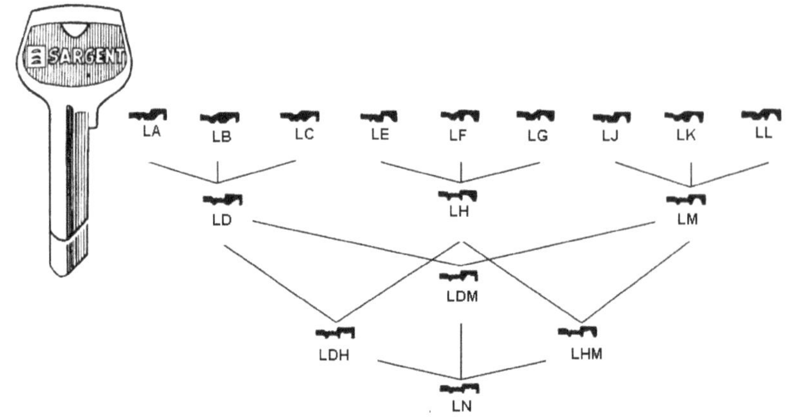

Sargent RA-RN

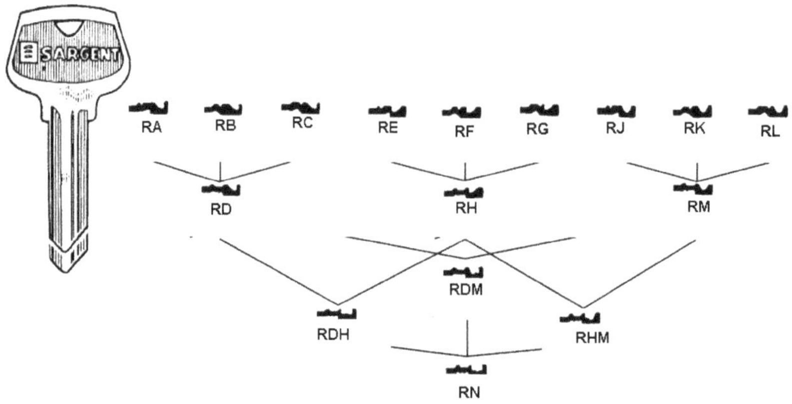

LESSON THIRTEEN
APPLICATION

If you have been following the book, you should now know how to determine your customer's needs and lay out a flow chart. You know how to progress a set of key bittings, master key bittings of various levels, etc. BUT HOW DO YOU TIE THE TWO TOGETHER?

It's really easier than you might think. The key to nearly everything is the Flow Chart you developed earlier.

Including any extra keys needed to allow for expansion, how many change keys fall under the lowest level of master key in the customer's system?

Based on the numbering system, method and format that you built your system on, what is the lowest level of Master Key that is capable of controlling that many change keys? That is the level you should give it. Assign that master key from any of those of that level on your chart, and go to other masters at the "same" level on the flow chart. Allowing for any future expansion that might be reasonably anticipated, determine the level of master key of each "MASTER" on the Flow Chart. They may vary in "Levels" even though they appear on the flow chart at the same "level".

Then go to the "GRAND MASTERS" on the Flow Chart. This time, the number of change keys below them is meaningless. Instead, look at the size of Master Key groups under it.

What level of Master Key will it take to control that many lower level master keys? Once again, we want to use as low a level as is possible, while still allowing for any future expansion needs.

We continue this way up the Flow Chart until we have it all filled out. The only thing remaining will be to determine the "best" way to accomplish special keying needs such as cross-keyed locks and single keyed locks.

Remember: Always work from the bottom UP when assigning levels.

A few additional guidelines:

- **If you can walk from one area to another and the same purpose is intended for the two areas, they will probably be keyed to the same change key.**
- **If many different people need to keyed access a low security area, it might be wise to leave some chambers empty in that cylinder.**
- **DO NOT assign bittings sequentially from the chart! Once you know the level select bittings for change keys randomly from within to make the system harder to predict.**

Lesson Fourteen:
KEYING INTO EXISTING SYSTEMS

Frequently the need arises to add new locks without changing the existing Master Key. Obviously we do not want any of these keys to operate existing locks, nor any of the existing keys to operate the new locks. This can be done, but it requires care on the part of the Master keying Locksmith.

Of course, the IDEAL situation is to have a healthy master key system in which proper keying records have been maintained (usually in a safe) and where no interchange now exists in the system. But many times that is quite simply not the case.

If interchange IS ALREADY NOW OCCURRING, it is time to re-do the system FROM SCRATCH. Many large institutions have farmed out the job of adding change keys to the system to locksmiths with neither the records of the job nor the necessary knowledge of proper master keying techniques. Any locks that have been done this way are not only a problem to the existing system, but will also affect any you now add to it.

But if interchange is NOT occurring, and the key records have still not been maintained, there is still hope for the system.

First, determine how the system was developed. What options were used? This can be done in much the same way as you did many of the projects throughout this book.

If Total Position Progression were used to develop it, it will be necessary to chase down and record each key already existing in the system, and compare it with the combinations on a chart built using this method. The remaining combinations are then available. This is a big job, but there is no easier way to do it and do it properly.

If Rotating Constant were used to develop the system, once again only that method can be used to expand it. Compare existing key bittings with the RCM Progression Patterns, and then expand it by using unused patterns. Once again, this is usually a major job.

But many systems are developed using Limited Progression technique. And the good news is, it can be easily expanded by considering the progressed positions as though they were part of a Rotating Constant system. Substitute P for each position progressed, and you now have one pattern from Rotating Constant.

For example, a system developed using the Limited Progression technique using three columns of progression, and using the Odometer (right to left) Sequence of Progression will have progressed positions in the three columns at the tip of the key. So will the system developed in a Holding 3 of 6 pattern of CCC**PPP**. They are identical.

So to expand the system, we simply use the remaining progression patterns to complement the one chosen, and build our new combinations. There will be NO INTERCHANGE. Of course, this is based on the existing system being healthy.

Although the rules describe the conditions by which a system is to be progressed, unless you have access to ALL the key bittings in a system, you cannot guarantee that the system was developed properly following the rule throughout in the past.

In fact, this is so important that the ALOA (Associated Locksmiths of America) put it into their Technical Standards Policy (ver. 2.3, 05/01/2005) the following:

11. Master Key System Integrity: No attempt will be made to expand any master keyed system without first obtaining a valid key bitting array and list of all key bittings currently in use. If this is not possible, the lock shall be keyed SKD or a new master key system generated.

Obviously we must be careful about any master keys of various levels in the system, since their capabilities may also be expanded.

We sometimes use the expanded capability when the master lower than the Top Master in the system (INCIDENTAL MASTERS) has already used all its combinations and now requires expansion. If the system is in Limited Progression, we simply expand the system using Rotating Constant, but choosing the part of a pattern that will give our Incidental Master more power.

The possibilities to do a professional job of master keying are bountiful. If you have kept up throughout the book, you should now be qualified to do virtually any system ever required of you.

One thing that MUST be stressed, however, is the importance of proper keying records and key control. Without these, the system will deteriorate to next to nothing.

Most professional Master keying Locksmiths also include a Key Storage Cabinet with each job, so that an original of each key can be kept on file AND NOT USED except to be duplicated from, preserving the proper operation of the locks and keys in the system.

Any records kept by the Locksmith should be kept in an encoded or Blind Coded state in a secure container or safe.

The supplement on the next page covers what factors must be considered in pricing or quoting a master keying job. Of course, if you work with institutional hardware, etc., your price is very likely to be your hourly wage.

But for the independent Master keying Locksmith, it is important that prices not be set on a whim or an arbitrary figure. And for them, the chart should help.

LESSON FIFTEEN:
PRICING

WALK TRAFFIC FLOW (system application layout)	HOURS x RATE = $
LAYOUT FLOW CHART & PROGRESSION CHART	HOURS x RATE = $
REMOVE CYLINDERS (number of cylinders x 8 mins.each) (OR ACTUAL TIME) Total mins. divided by 60 =total hrs.	HOURS x RATE = $
PINNING CYLINDERS (replacing all pins & springs) Number of cyls. x price per cyl.	CYLS. x PRICE = $
CHANGE KEYS CUT BY CODE	CKS x PRICE PER = $
KEYS DUPLICATED FOR CHANGE KEYS	DUPS x PRICE PER = $
MASTER KEYS CUT BY CODE (quantity x code cutting price)	MKS x PRICE PER = $
STAMP KEYS (STANDARD KEY CODING SYSTEM, symbols, #s, etc,)	KEYS x PRICE = $
RE-INSTALL CYLINDERS IN EXISTING LOCKS AND CHECK OPERATION. (Number of cyls. x 12 mins. ea. =total mins. Total mins. divided by 60 = hrs. (OR ACTUAL TIME)	HOURS x RATE = $
SERVICE CALL CHARGES	CALL x RATE = $
WALK-CHECK SYSTEM WITH AUTHORIZED PERSON IN CHARGE	x = $
KEY CABINET COST (quantity x price each)	EACH x PRICE = $

You should always include a key cabinet. If you do not, it encourages the client to issue too many master keys, resulting in earlier system breakdown and customer dissatisfaction. Also, whenever possible, provide some key management suggestions for their usage, or the co-creation of a Key Management Policy.

The following definitions are Master Keying relevant excerpts from the LIST Council Dictionary, which states:

" Copyright □ , 2005 by the Lock Industry Standards and Training Council.

All rights reserved.

Permission is hereby granted to reprint terms and definitions contained herein with the following stipulations:

1. Any term and its definition are to be reprinted in their entirety.
2. Credit is to be given to the Lock Industry Standards and Training (LIST) Council."

associated change key
n. a change key which is related directly to particular master key(s) through the use of constant cuts

associated master key
n. a master key which has particular change keys related directly to its combination through the use of constant cuts

attendant's key
n. a selective master key used in a hospital keying system

bible
n. that portion of the cylinder shell which normally houses the pin chambers, especially those of a key-in-knob cylinder or certain rim cylinders

bicentric cylinder
n. a cylinder which has two independent plugs, usually with different keyways. Both plugs are operable from the same face of the cylinder. It is designed for use in extensive master key systems.

bi-directional cylinder
n. a bitting
n. 1. the number(s) which represent(s) the dimensions of the key 2. the actual cut(s) or combination of a key

bitting depth
n. the depth of a cut which is made into the blade of a key

bitting list
n. a listing of all the key combinations used within a system. The combinations are usually arranged in order of the blind code, direct code, and/or key symbol.

bitting orientation
n. the specification of how a written combination is to be applied to bitting positions of a key (tip-to-bow, etc.)

bitting position
n. the location of a key cut

blade
n. the portion of a key which may contain the cuts and/or millings

blank
adj. uncut, see also "key blank"

blind code
n. a designation, unrelated to the bitting, assigned to a particular key combination for future reference when additional keys or cylinders may be needed

block master key
n. the one pin master key for all combinations listed as a block in the standard progression format cylinder which may be operated in a clockwise and counterclockwise direction by a single key

BMK
abb. block master key

bottom of blade
n. the portion of the blade opposite the cut edge of a single bitted key

bottom pin
n. usually a cylindrical shaped tumbler which may be conical, ball shaped or chisel pointed on the end which makes contact with the key

bow
n. the portion of the key which serves as a grip or handle

bow stop
n. a type of stop located near the key bow

bow to tip
n. a specification that a bitting combination is recorded starting at the bow and proceeding to the tip

broach
n. 1. a tool used to cut the keyway into the cylinder plug 2. v. to cut the keyway into a cylinder plug with a broach

building master key
n. a master key which operates all or most master keyed looks in a given building

build-up dimension
n. 1. the distance between two different shear lines, as expressed in units of the manufacturer's increment or as an actual measurement 2. the dimension of the buildup pin required in a particular chamber, which will allow one key to operate at the plug shear line and a different key to operate at a different shear line

build-up pin
n. the additional element of a pin stack required to allow operation at different shear lines in a cylinder

cap
1. n. a spring cover for a single pin chamber 2. n. a part which may serve as a plug retainer and/or a holder for the tailpiece 3. v. to install a cap

capping block

n. a holding fixture for certain interchangeable cores which aids in the installation of the caps

carriage
n. 1. the movable assembly of a key machine which supports the key vise(s), 2. a changeable part of a clipper that holds the key blank

CBOK
abb. construction breakout key

chamber
n. any cavity in a cylinder plug and/or shell which houses the tumblers

change key
n. 1. a key which operates only one cylinder or one group of keyed alike cylinders in a keying system, 2. any device that is used to mechanically or electronically allow resetting of certain key or combination locks, see also "reset key" #1

check pin
n. a device resembling a bottom pin, found in a non-traditional chamber, used to restrict the rotation of the plug unless a specific area of the key is cut

CK
1. abb. change key 2. abb. control key

clipper
n. a hand held key bitting punch, often incorporating a trigger-like handle

CMK
abb. construction master key

CMK'd
abb. construction master keyed

code
n. 1. a designation assigned to a particular key combination for reference when additional keys or cylinders may be needed. See also, "blind code", "direct code", and "key symbol", 2. a standard for Life Safety, structural integrity, fire safety, etc., either incorporated into the statutes of a given jurisdiction, or a model standard

code key
n. a key cut to a specific code rather than duplicated from a pattern key. it may or may not conform to the lock manufacturer's specifications

code original key
n. a code key which conforms to the lock manufacturer's specifications

code series
n. an ordered sequence of unique codes related to specific key combinations, key sections and manufacturers

composite blank
n. a blank designed to enter two or more simplex keyways

composite keyway
n. a keyway which has been enlarged to accept more than one key section, often key sections of more than one manufacturer

constant cut
n. any bitting(s) which are identical in corresponding positions from one key to another in a keying system. They usually serve to group these keys together within a given level of keying, and/or link them with keys of other levels. See also "rotating constant"

construction breakout key
n. a key used by some manufacturers to render all construction master keys permanently inoperative

construction core
n. an interchangeable or removable core designed for use during the construction phase of a building. The cores are normally keyed alike and, upon completion of construction, they are to be replaced by the permanent system's cores.

construction master key
n. a key normally used by construction personnel for a temporary period during building construction. It may be rendered permanently inoperative without disassembling the cylinder.

construction master keyed
adj. of or pertaining to a cylinder which is or is to be operated temporarily by a construction master key

control chamber
n. in an interchangeable or removable core, any chamber which has a control shear line, which is different from the operating shear line

control cut
n. any bitting which operates the retaining device of an interchangeable or removable core

control dimension
n. in certain interchangeable or removable cores, the distance between the operating shear line and the control shear line, expressed either in units of the increment or as a measurement

control key
n. 1. a key whose only purpose is to remove and/or install an interchangeable or removable core 2. a bypass key used to operate and/or reset some combination type locks 3. a key which allows disassembly of some removable cylinder locks

control lug
n. that part of an interchangeable or removable core-retaining device which locks the core into its housing

control number
n. a number used to calculate the size of the build-up pin when; (a) the sizes of bottom and master pins are known, or (b) the bitting depths of all operating keys are known

control shear line
n. the shear line which allows operation of the control lug of an interchangeable or removable core

control sleeve
n. the part of an interchangeable or removable core retaining device which surrounds the plug

controlled cross keying
n. a condition in which two or more different keys of the same level of keying and under the same higher level key(s) operate one cylinder by design; e.g., XAA1 operated by AA2 (but not XAA1 operated by AB1)
Note: This condition could severely limit the security of the cylinder and the maximum expansion of the system when (1) more than a few of these different keys operate a cylinder, or (2) more than a few differently cross keyed cylinders per system are required.

core
n. a complete unit, often with a "figure eight" shape, which usually consists of the plug, shell, tumblers, springs, plug retainer and spring cover(s). It is primarily used in removable and interchangeable core cylinders and looks.

cross keying
n. the deliberate process of combinating a cylinder (usually in a master key system) to two or more different keys which would not normally be expected to operate it together. See also "controlled cross keying" and "uncontrolled cross keying."

cut key
n. a key which has been bitted or combinated

cut root
n. the bottom of a key cut

cut root shape
n. the shape of the bottom of a key cut. It may have a flat or radius of a specific dimension, or be a perfect "V"

cutter

n. the part of a key machine which makes the cuts into the key blank

cutter guard
n. a shield over a key machine cutter to protect the operator

cylinder
n. a complete operating unit which usually consists of the plug shell, tumblers, springs, plug retainer, a cam/tailpiece or other actuating device, and all other necessary operating parts

cylinder blank
n. a dummy cylinder which has a solid face and no operating parts

cylinder clip
n. a spring steel device used to secure some types of cylinders

cylinder key
n. a broad generic term including virtually all keys for pin and disc tumbler locks

cylinder lock
n. any lock that is operated by a cylinder key

declining step key
n. a key whose cuts are equal or deeper as you proceed from bow to tip

department master key
n. a master key which operates all or most master keyed locks of a given department

depth key set
n. a set of keys used to make a code original key on a key duplicating machine to a lock manufacturer's given set of key bitting specifications. Each key is cut with the correct spacing to one depth only in all bitting positions, with one key for each depth.

derived series
n. a series of blind codes and bittings which are directly related to those of another bitting list

direct code
n. a designation assigned to a particular key which includes the actual combination of the key

display key
n. a special change key in a hotel master key system which will allow access to one designated guest room, even if the lock is in the shut out mode. It may also act as a shut out key for that room.

driver spring
n. a spring placed on top of the pin stack to exert pressure on the pin tumblers

duplicate key
n. any key reproduced from a pattern key

effective plug diameter
n. the dimension obtained by adding the root depth of a key cut to the length of its corresponding bottom pin which establishes a perfect shear line. This will not necessarily be the same as the actual plug diameter.

emergency master key
n. a special master key which usually operates all guestroom locks in a hotel master key system at all times, even in the shut out mode. This key may also act as a shut out key.

EMK
abb. emergency master key

ENG
abb. engineer's key

engineer's key
n. a selective master key which is used by maintenance personnel to operate many locks under different master keys in a system of three or more levels of keying

expansion specification
n. a requirement for bittings to be planned beyond the immediate needs of a keying system

factory original key
n. the cut key furnished by the lock manufacturer for a lock or cylinder

first key
n. any key produced without the use of a pattern key

five column progression
n. a process wherein key bittings are obtained by using the cut possibilities in five columns of the key bitting array

five pin master key
n. a master key for all combinations obtained by progressing five bitting positions

floor master key
n. a master key which operates all or most master keyed locks on a particular floor of a building

four column progression
n. a process wherein key bittings are obtained by using the cut possibilities in four columns of the key bitting array

four pin master key
n. a master key for all combinations obtained by progressing four bitting positions

GGGMK
abb. great great grand master key

GGGMK'D
abb. great great grand master keyed

GGM
abb. great grand master key

GGMK
abb. great grand master key

GGMK'D
abb. great grand master keyed

GM
abb. grand master key

GMK
abb. grand master key

GMK section
abb. grand master key section

GMK'D
abb. grand master keyed

graduated drivers
n. a set of top pins of different lengths. Usage is based on the height of the rest of the pin stack, in order to achieve a uniform pin stack height. See also "compensate drivers"

grand master key
n. the key which operates two or more separate groups of locks, which are each operated by a different master key

grand master key system
n. a master key system which has exactly three levels of keying

grand master keyed
adj. of or pertaining to a lock or cylinder which is or is to be keyed into a grand master key system

great grand master key
n. the key which operates two or more separate groups of locks which are each operated by a different grand master key

great grand master key system
n. a master key system which has exactly four levels of keying

great grand master keyed
adj. of or pertaining to a lock or cylinder which is or is to be keyed into a great grand master key system

great great grand master key

n. the key which operates two or more separate groups of locks which are each operated by different great grand master keys

great great grand master key system
n. a master key system which has five or more levels of keying

great great grand master keyed
adj. of or pertaining to a lock or cylinder which is or is to be keyed into a great great grand master key system

guest key
n. a key in a hotel master key system which is normally used to unlock only the one guest room for which it was intended, but will not operate the lock in the shut out mode

guide
n. 1. that part of a key machine which follows the cuts of a pattern key or template during duplication 2. that part of a flat key lever lock which connects the nose to the foot and supports the key blade

HGM

abb. horizontal group master key

high security cylinder
n. a cylinder which offers a greater degree of resistance to any two or more of the following: picking, impressioning, key duplication, drilling or other forms of forcible entry

high security key
n. a key for a high security cylinder

HKP
abb. housekeeper's key

horizontal group master key
n. the two pin master key for all combinations listed in all blocks in a line across the page in the standard progression format

hotel cylinder
n. a cylinder which can function to prevent the operation of all keys except emergency, display and some shut out keys. It may also contain an indicator.

housekeeper's key
n. a selective master key in a hotel master key system which may operate all guest and linen rooms and other housekeeping areas

housing
n. that part of a locking device which is designed to hold a core

IC
abb. interchangeable core

incidental master key

n. a key cut to an unplanned shear line created when the cylinder is combinated to the top master key and a change key

increment
n. a usually uniform increase or decrease in the successive depths of a key cut which must be matched by a corresponding change in the tumblers

individual key
n. an operating key for a lock or cylinder which is not part of a keying system, see also "change key" #1

interchangeable core
n. a key removable core which can be used in all or most of the core manufacturer's product line. No tools (other than the control key) are required for removal of the core.

jumbo cylinder
n. a rim or mortise cylinder of 1-1/2" diameter

k
sym. symbol for "keys" used after a numerical designation of the quantity of the keys requested to be supplied with the cylinders; e.g., lk, 2k, 3k, etc. it is usually found in hardware/keying schedules.

KA
abb. keyed alike

KA1, KA2, etc.
symbol which indicates that all cylinders so designated are or are to be operated by the same key(s). The numerical designation indicates the keyed alike group or set.

KA/2, KA/3, etc.
symbol used to indicate the quantity of locks or cylinders in keyed alike groups. These groups are usually formed from a larger quantity; e.g., 30 cylinders KA/2.

KBA
abb. key bitting array

KD
abb. keyed different

key
n. a properly combinated device which is, or most closely resembles, the device specifically intended by the lock manufacturer to operate the corresponding lock

key bitting array
n. a matrix (graphic) display of all possible bittings for change keys and master keys as related to the top master key

key bitting punch
n. a manually operated device which stamps or punches the cuts into the key blade, rather than grinding or milling them

key bitting specifications
n. pl. the technical data required to bit a given (family of) key blank(s) to the lock manufacturer's dimensions

key blank
n. any material manufactured to the proper size and configuration which allows its entry into the keyway of a specific locking device. A key blank has not yet been combinated or cut.

key cabinet
n. a cabinet with hooks, velcro, or other means designed to store keys systematically

key caliper
n. an outside caliper specifically made with an anvil of size and shape to measure the root depth of bittings

key control
n. 1. any method or procedure which limits unauthorized acquisition of a key and/or controls distribution of authorized keys 2. a systematic organization of keys and key records

key cut(s)
n. the portion of the key blade which remains after being cut and which aligns the tumbler(s)

key cut profile
n. the shape of a key cut, including the cut angle and the out root shape

key duplicating machine
n. a key machine which is designed to make copies from a pattern key

key gauge
n. a usually flat device with a cutaway portion indexed with a given set of depth or spacing specifications. It is used to help determine the combination of a key.

key holder
n. a person to whom a key has been issued

key interchange
n. an undesirable condition, usually in a master key system, whereby a key unintentionally operates a cylinder or lock

key records
n. pl. records which typically include some or all of the following: bitting list, key bitting array, key system schematic, end user, number of keys/cylinders issued, names of persons to whom keys were issued, hardware/keying schedule

Key Records Department
n. the department which is responsible for generating and issuing all lock and key combinations and maintaining records of them

key rest
n. a non-movable part of a key machine vise that is used to position and align the bottom of the blade

key section
n. the exact cross sectional configuration of a key blade as viewed from the bow toward the tip

key section identifier
n. any tool, device or software used to identify a particular key section

key symbol
n. a designation used for a key combination in the standard key coding system, e.g., A, AA, AA1, etc.

key system schematic
n. a drawing with blocks utilizing keying symbols, usually illustrating the hierarchy of all keys within a master key system. It indicates the structure and total expansion of the system.

keyed alike
adj. of or pertaining to two or more looks or cylinders which have or are to have the same combination. They may or may not be part of a keying system

keyed different
adj. of or pertaining to a group of locks or cylinders, each of which is or is to be combinated differently from the others. They may or may not be part of a keying system.

keyed random
adj. of or pertaining to a cylinder or group of cylinders selected from a limited inventory of different key changes. Duplicate bittings may occur.

keying
n. any specification for how a cylinder or group of cylinders are or are to be combinated in order to control access

keying conference
n. a meeting of the and user and the keying system supplier at which the keying and levels of keying, including future expansion, are determined and specified

keying kit
n. a compartmented container which holds an assortment of tumblers, springs and/or other parts

keying schedule
n. a detailed specification of the keying system listing how all cylinders are to be keyed and the quantities, markings, and shipping instructions of all keys and/or cylinders to be provided

keying symbol

n. a designation used for a lock or cylinder combination in the standard key coding system; e.g., AA1, XAA1, XIX, etc.

keyway
n. 1. the opening in a lock or cylinder which is shaped to accept key bit or blade of a proper configuration 2. the exact cross sectional configuration of a keyway as viewed from the front. It is not necessarily the same as the key section.

KWY
abb. keyway

large format interchangeable core
n. an interchangeable core which is too large to fit into a small format interchangeable core housing

levels of keying
n. pl. the divisions of a master key system into hierarchies of access, as shown in the following tables. Note: the standard key coding system has been expanded to include key symbols for systems of more than four levels of keying.

TWO LEVEL SYSTEM

LEVEL OF KEYING	KEY NAME	ABB.	KEY SYMBOL
Level II	master key	MK	AA
Level I	change key	CK	1AA, 2AA, etc.

THREE LEVEL SYSTEM

LEVEL OF KEYING	KEY NAME	ABB.	KEY SYMBOL
Level III	grand master key	GMK	A
Level II	master key	MK	AA, AB, etc.
Level I	change key	CK	AA1, AA2, etc

FOUR LEVEL SYSTEM

LEVEL OF KEYING	KEY NAME	ABB.	KEY SYMBOL
Level IV	great grand master key	GGMK	GGMK
Level III	grand master key	GMK	A, B, etc.
Level II	master key	MK	AA, AB, etc.
Level I	change key	CK	AA1, AA2, etc.

FIVE LEVEL SYSTEM

LEVEL OF KEYING	KEY NAME	ABB.	KEY SYMBOL
Level V	great great grand master key	GGGMK	GGGMK
Level IV	great grand master key	GGMK	A, B, etc.
Level III	grand master key	GMK	AA, AB, etc.
Level II	master key	MK	AAA, AAB, etc.
Level I	change key	CK	AAA1, AAA2, etc.

SIX LEVEL SYSTEM

LEVEL OF KEYING	KEY NAME	ABB.	KEY SYMBOL
Level VI	great great grand master key	GGGMK	GGGMK
Level V	great grand master key	GGMK	A, B, etc.
Level IV	grand master key	GMK	AA, AB, etc.
Level III	master key	MK	AAA, AAB, etc.
Level II	sub-master key	SMK	AAAA, AAAB, etc.
Level I	change key	CK	AAAA1, AAAA2, etc.

LFIC
abb. Large Format Interchangeable Core

lost ball keying
n. a construction keying method that uses a tiny ball(s) in the pin stack to allow construction master key operation. The ball(s) may be removed from the pin stack without disassembly of the cylinder.

MACS
abb. maximum adjacent cut specification

maid's master key
n. the master key in a hotel master key system given to the maid. It operates only cylinders of the guestrooms and linen closets in the maid's designated area.

maison key system
n. [from the French, meaning "house" key system] a keying system in which one or more cylinders are operated by every key (or relatively large numbers of different keys) in the system; e.g., main entrances of apartment buildings operated by all individual suite keys of the building

master blank
n. a key blank used in a complementary keyway lock, on which only the master key combination is intended to be cut

master key
n. 1. a key which operates all the master keyed locks or cylinders in a group, each lock or cylinder usually operated by its own change key, 2. v. to combinate a group of locks or cylinders such that each is operated by its own change key as well as by a master key for the entire group, 3. n. an automotive key which operates all or most locks on a vehicle where there is also a valet key employed

master key changes
n. the number of different usable change keys available under a given master key

master key system
n. 1. any keying arrangement which has two or more levels of keying 2. a keying arrangement which has exactly two levels of keying

master keyed
adj. 1. of or pertaining to a cylinder or group of cylinders which are or are to be combinated so that all may be operated by their own change key(s) and by additional key(s) known as master key(s) 2. a warded lock whose warding has been designed to allow different hierarchies of keys to operate

master keyed only
adj. of or pertaining to a lock or cylinder which is or is to be combinated only to a master key

master pin
n. 1. usually a cylindrical shaped tumbler, flat on both ends, placed between the top and bottom pin to create an additional shear line 2. a pin tumbler with multiple gates to accept a sidebar

master ring
n. a tube shaped sleeve located between the plug and shell of certain cylinders to create a second shear line. Normally the plug shear line is used for change key combinations and the shell shear line is used for master key combinations.

master ring lock/cylinder
n. a lock or cylinder equipped with a master ring

matrix format master keying
n. a method of generating combinations and assigning key symbols in a master key system through the use of a matrix

maximum adjacent cut specification
n. the maximum allowable difference between adjacent cut depths

maximum compound cut specification
n. the maximum allowable difference between the different depths of a compound cut

maximum opposing cut specification
n. the maximum allowable depths to which opposing cuts can be made without breaking through the key blade. This is typically a consideration with dimple keys.

MCCS
Abb. Maximum Compound Cut Specification

MK
abb. master key

MKCS
abb. modified key coding system

MK'd
abb. master keyed

MK'd only
abb. master keyed only

MK section
abb. master key section

MOCS
abb. maximum opposing cut specification

modified key coding system
n. a modification of the standard key coding system, in which the TMK of a system with three or more levels of keying has a single letter name, and keys at lower levels are named following SKCS conventions

multiplex key blank
n. any key blank which is part of a multiplex key system

multiplex key system
n. 1. a series of different key sections which may be used to expand a master key system by repeating bittings on additional key sections. The keys of one key section will not enter the keyway of another key section. This type of system always includes another key section which will enter more than one, or all of the keyways. 2. a keying system which uses such keyways and key sections

mushroom pin
n. a pin tumbler, usually a top pin, which resembles a mushroom. It is typically used to increase pick resistance

mykey chart
n. a matrix depicting user requested cross keying by listing keys along one axis and openings along the other. So named because it is frequently accompanied by the statement "My key goes here and here and here."

NCK
sym. symbol for "no change key," primarily used in hardware schedules

NMK
abb. a notation used to indicate "not master keyed" and is suffixed in parentheses to a regular keying symbol. It indicates that the cylinder is not to be operated by the master key(s) specified in the regular keying symbol; e.g., AB6(NMK).

nurse's key
n. a selective master key used in a hospital keying system

odometer method
n. a means of progressing key bittings using a progression sequence of right to left.

one column progression
n. a process wherein key bittings are obtained by using the cut possibilities in one column of the key bitting array

one pin master key
n. a master key for all combinations obtained by progressing only one bitting position

operating key
n. any key which will properly operate a lock or cylinder to lock or unlock the lock mechanism and is not a control key or reset key, see also "change key"

operating shear line
n. any shear line which allows normal operation of a cylinder or lock

original key blank
n. a key blank supplied by the lock manufacturer to fit that manufacturer's specific product

page master key
n. the three-pin master key for all combinations listed on a page in the standard progression format

paracentric
adj. 1. of or pertaining to a keyway with one or more wards on each side projecting beyond the vertical center line of the keyway to hinder picking 2. of or pertaining to a key blank made to enter such a keyway

parity
n. the property of being either even or odd, as the depth number of a bitting in a combination, or the possible progressives in a given position of a two step progression

parity pattern
n. the collective description of the parities of a group of bitting positions in a two step progression, typically expressed in an even/odd sequence, e.g. EOOEEO

patented unrestricted keyway
n. a specific keyway which is sold without restriction by a lock maker, but which retains patent protection against non-original key blank makers

pattern key
n. 1. an original key kept on file to use in a key duplicating machine when additional keys are required, 2. any key which is used in a key duplicating machine to create a duplicate key

pin
v. to install pin tumblers into a cylinder and/or cylinder plug, see also "pin tumbler"

pin chamber
n. the corresponding hole drilled into the cylinder shell and/or plug to accept the pin(s) and spring

pin kit
n. a type of keying kit for a pin tumbler mechanism

pin stack
n. all the tumblers in a given pin chamber, see also"pin stack height"

pin stack height
n. the measurement of a pin stack, often expressed in units of the lock manufacturer's increment or as an actual dimension

pin tray
n. see "layout tray"

pin tumbler
n. usually a cylindrical shaped tumbler. Three types are normally used: bottom pin, master pin and top pin.

pin tweezers
n. pl. a tool used in handling tumblers and springs

pinning block
n. a holding fixture which assists in the loading of tumblers into a cylinder or cylinder plug

pinning chart
n. a numerical diagram which indicates the sizes and order of installation of the various pins into a cylinder. The sizes are usually indicated by a manufacturer's reference number which equals the quantity of increments a tumbler represents.

plug
n. the part of a cylinder which contains the keyway, with tumbler chambers usually corresponding to those in the cylinder shell

plug follower
n. a tool used to allow removal of the cylinder plug while retaining the top pins, springs, and/or other components within the shell

plug holder
n. a holding fixture which assists in the loading of tumblers into a cylinder plug

plug retainer
n. the cylinder component which secures the plug in the shell

positional master keying
n. a method of master keying typical of certain binary type disc tumbler key-in-knob looks and of magnetic and dimple key cylinders. Of all possible tumbler positions within a cylinder, only a limited number contain active tumblers. The locations of these active tumblers are rotated among all possible positions to generate key changes. Higher level keys must have more cuts or magnets than lower level keys.

practical key changes
n. pl. the total number of usable different combinations available for a specific cylinder or lock mechanism

progress
v. to select possible key bittings, usually in numerical order, from the key bitting array

progression
n. a logical sequence of selecting possible key bittings, usually in numerical order from the key bitting array. 2. a system of generating keys with different cut combinations for unknown positions, when a partial bitting is known

progression chart
n. a list of combinations to consider when fitting keys to a lock with only a partial bitting known, typically arranged to minimize the number of key blanks needed

progression column
n. a listing of the key bitting possibilities available in one bitting position as displayed in a column of the key bitting array

progression list
n. a bitting list of change keys and master keys arranged in sequence of progression

progressive
n. any bitting position which is progressed rather than held constant

proprietary
adj. of or pertaining to a keyway and key section assigned exclusively to one end user by the lock manufacturer. It may also be protected by law from duplication.

protected key
n. a physical key that affords legal remedies against duplication

random master keying

n. any undesirable process used to master key which uses unrelated keys to create a system

recombinate
v. to change the combination of a lock, cylinder or key

recore
v. to rekey by installing a different core

register groove
n. the reference point on the key blade from which some manufacturers index the bitting depths

register number
n. 1. a reference number, typically assigned by the lock manufacturer to an entire master key system 2. a blind code assigned by some lock manufacturers to higher level keys in a master key system

rekey
v. to change the existing combination of a cylinder or lock

removable core
n. a key removable core which can only be installed in one type of cylinder housing; e.g., rim cylinder or mortise cylinder or key-in-knob lock

removable cylinder
n. a cylinder which can be removed from a locking device by a key and/or tool

repin
v. to replace pin tumblers, with or without changing the existing combination

restricted
adj. of or pertaining to a keyway and corresponding key blank whose sale and/or distribution is limited by the lock manufacturer in order to reduce unauthorized key proliferation

reverse keyway
n. a keyway shape that is a mirror image of another

rim cylinder
n. a cylinder typically used with surface applied locks and attached with a back plate and machine screws. It has a tailpiece to actuate the lock mechanism

rim exit device
n. an exit device mounted on the door surface

rim lock
n. a lock or latch typically mounted on the surface of a door or drawer

RM
abb. row master key

rotating constant
n. one or more cut(s) in a key of any level which remain constant throughout all levels and are identical to the top master key cuts in their corresponding positrons. The positions where the top master key cuts are held constant may be moved, always in a logical sequence

rotating constant method
n. a method used to progress key bittings in a master key system, wherein at least one cut in each key is identical to the corresponding cut in the top master key. The identical cut(s) is moved to different locations in a logical sequence until each possible planned position has been used.

row master key
n. the one-pin master key for all combinations listed on the same line across a page in the standard progression format

selective key system
n. a key system in which every key has the capability of being a master key. It is normally used for applications requiring a limited number of keys and extensive cross keying.

selective master key
n. an unassociated master key which can be made to operate any specific lock(s) in the entire system in addition to the regular master key(s) and/or change key(s) for the cylinder without creating key interchange

sequence of progression
n. the order in which bitting positions are progressed to obtain change key combinations

set-up plug
n. a type of loading tool shaped like a plug follower. It contains pin chambers and is used with a shove knife to load springs and top pins into a cylinder shell.

seven column progression
n. a process wherein key bittings are obtained by using the cut possibilities in seven columns of the key bitting array

seven pin master key
n. a master key for all combinations obtained by progressing seven bitting positions

SFIC
abb. Small Format Interchangeable Core

shear line
n. a location in a cylinder at which specific tumbler surfaces must be aligned, removing obstruction(s) which prevented the plug from moving

shell
n. the part of the cylinder which surrounds the plug and which usually contains tumbler chambers corresponding to those in the plug

shoulder
n. any key stop other than a tip stop, see also "bow stop"

shut out key
n. usually used in hotel keying systems, a key which will make the lock inoperative to all other keys in the system except the emergency master key, display key, and some types of shut out keys

shut out mode
n. the state of a hotel function lockset which prevents operation by all keys except the emergency master key, display key, and some types of shut out keys

simplex key section
n. a single independent key section which cannot be used in a multiplex key system

single key section
n. an individual key section which can be used in a multiplex key system

single step progression
n. a progression using a one increment difference between bittings of a given position

six column progression
n. a process wherein key bittings are obtained by using the cut possibilities in six columns of the key bitting array

six pin master key
n. a master key for all combinations obtained by progressing six bitting positions

SKCS
abb. standard key coding system

SKD
sym. symbol for "single keyed", normally followed by a numerical designation in the standard key coding system; e.g., SKD1, SKD2, etc. It indicates that a cylinder or lock is not master keyed but is part of the keying system

Small Format Interchangeable Core
n. an IC that replicates the functionality and design popularized by Best

SMK
abb. sub-master key

spacing
n. the dimensions from the stop to the center of the first cut and/or to the centers of successive cuts

split pin master keying
n. a method of master keying a pin tumbler cylinder by installing master pins into one or more pin chambers

spool pin

n. a pin with a uniform smaller diameter except at its ends. Typically a top pin used to augment pick resistance but also observed as bottom and master pins.

spring cover
n. a device for sealing one or more pin chambers

standard key coding system
n. an industry standard and uniform method of designating all keys and/or cylinders in a master key system. The designation automatically indicates the exact function and keying level of each key and/or cylinder in the system, usually without further explanation.

standard progression format
n. a systematic method of listing and relating all change key combinations to all master key combinations in a master key system. The listing is divided into segments known as blocks, horizontal groups, vertical groups, rows, and pages, for levels of control.

step pin
n. a spool or mushroom pin which has had a portion of its end machined to a smaller diameter than the opposite end. It is typically used as a top pin to improve pick resistance by some manufacturers of high security cylinders.

stop (of a key)
n. the part of a key from which all cuts are indexed and which determines how far the key enters the keyway

sub-master key
n. the master key level immediately below the master key in a system of six or more levels of keying

substitution code
n. a code whose individual characters are converted to individual key cuts or combination numbers by means of a reference table

tumbler
n. a movable obstruction of varying size and configuration in a lock or cylinder which makes direct contact with the key or another tumbler and prevents an incorrect key or torque device from activating the lock or other mechanism

tumbler spring
n. any spring which acts directly on a tumbler

two-step progression
n. a progression using a two increment difference between bittings of a given position

two pin master key
n. a master key for all combinations obtained by progressing two bitting positions

two column progression
n. a process wherein key bittings are obtained by using the cut possibilities in two columns of the key bitting array

top master key
n. the highest level master key in a master key system

top of blade
n. the bitted edge of a single bitted key

top pin
n. usually a cylindrical shaped tumbler, usually flat on both ends and installed directly under the spring in the pin stack

total position progression
n. a process used to obtain key bittings in a master key system wherein bittings of change keys differ from those of the top master key in all bitting positions

TMK
abb. top master key

tip
n. the portion of the key which enters the keyway first

tip stop
n. a type of stop located at or near the tip of the key

tip to bow
n. a specification that a bitting combination is recorded starting at the tip and proceeding to the bow

three pin master key
n. a master key for all combinations obtained by progressing three bitting positions

three column progression
n. a process wherein key bittings are obtained by using the cut possibilities in three columns of the key bitting array

theoretical key changes
n. pl. the total possible number of different combinations available for a specific cylinder or lock mechanism

unassociated change key
n. a change key which is not related directly to a particular master key through the use of certain constant cuts

unassociated master key
n. a master key which does not have change keys related to its combination through the use of constant cuts

uncombinated
adj. 1. of or pertaining to a cylinder which is or is to be supplied without keys, tumblers and springs 2. of or pertaining to a lock, cylinder or key in which the combination has not been set

uncontrolled cross keying

n. a condition in which two or more different keys under different higher level keys operate one cylinder by design; e.g., XAA1 operated by AB, AB1. Note: This condition severely limits the security of the cylinder and the maximum expansion of the system, and often leads to key interchange

visual key control
n. a specification that all keys and the visible portion of the front of all lock cylinders be stamped with standard keying symbols

vertical group master key
n. the two-pin master key for all combinations listed in all blocks in a line down a page in the standard progression format

X
sym. symbol used in hardware schedules to indicate a cross-keyed condition for a particular cylinder; e.g., XAA2, XIX (but not AX7)

THEORETICAL MASTER

(KBA)
KEY BITTING ARRAY

PAGE MASTER ______
GROUP MASTER ______
SECTION MASTER ______

PAGE # ______
OF ______
PAGES

⇨ SEQUENCE OF PROGRESSION

LINE MASTERS

RIZONTAL MASTER

VERTICAL MASTER — 1, 2, 3, 4
VERTICAL MASTER — 17, 18, 19, 20
VERTICAL MASTER — 33, 34, 35, 36
VERTICAL MASTER — 49, 50, 51, 52

ZONTAL MASTER

BLOCKMASTER — 5, 6, 7, 8
BLOCKMASTER — 21, 22, 23, 24
BLOCKMASTER — 37, 38, 39, 40
BLOCKMASTER — 53, 54, 55, 56

ZONTAL MASTER

BLOCKMASTER — 9, 10, 11, 12
BLOCKMASTER — 25, 26, 27, 28
BLOCKMASTER — 41, 42, 43, 44
BLOCKMASTER — 57, 58, 59, 60

RIZONTAL MASTER

BLOCKMASTER — 13, 14, 15, 16
BLOCKMASTER — 29, 30, 31, 32
BLOCKMASTER — 45, 46, 47, 48
BLOCKMASTER — 61, 62, 63, 64

BLOCKMASTER
BLOCKMASTER
BLOCKMASTER
BLOCKMASTER

THEORETICAL MASTER

(KBA)
KEY BITTING ARRAY

PAGE MASTER ______
GROUP MASTER ______
SECTION MASTER ______

PAGE # ______
OF ______
PAGES

⇦ SEQUENCE OF PROGRESSION

LINE MASTERS

RIZONTAL MASTER

VERTICAL MASTER
1
2
3
4

VERTICAL MASTER
17
18
19
20

VERTICAL MASTER
33
34
35
36

VERTICAL MASTER
49
50
51
52

ZONTAL MASTER

BLOCKMASTER
5
6
7
8

BLOCKMASTER
21
22
23
24

BLOCKMASTER
37
38
39
40

BLOCKMASTER
53
54
55
56

ZONTAL MASTER

BLOCKMASTER
9
10
11
12

BLOCKMASTER
25
26
27
28

BLOCKMASTER
41
42
43
44

BLOCKMASTER
57
58
59
60

RIZONTAL MASTER

BLOCKMASTER
13
14
15
16

BLOCKMASTER
29
30
31
32

BLOCKMASTER
45
46
47
48

BLOCKMASTER
61
62
63
64

BLOCKMASTER

BLOCKMASTER

BLOCKMASTER

BLOCKMASTER

THEORETICAL MASTER

(KBA)
KEY BITTING ARRAY

PAGE MASTER
GROUP MASTER
SECTION MASTER

PAGE #
OF
PAGES

SEQUENCE OF PROGRESSION

LINE MASTERS

VERTICAL MASTER | VERTICAL MASTER | VERTICAL MASTER | VERTICAL MASTER

RIZONTAL MASTER
1 2 3 4 | 17 18 19 20 | 33 34 35 36 | 49 50 51 52

BLOCKMASTER | BLOCKMASTER | BLOCKMASTER | BLOCKMASTER

ZONTAL MASTER
5 6 7 8 | 21 22 23 24 | 37 38 39 40 | 53 54 55 56

BLOCKMASTER | BLOCKMASTER | BLOCKMASTER | BLOCKMASTER

ZONTAL MASTER
9 10 11 12 | 25 26 27 28 | 41 42 43 44 | 57 58 59 60

BLOCKMASTER | BLOCKMASTER | BLOCKMASTER | BLOCKMASTER

RIZONTAL MASTER
13 14 15 16 | 29 30 31 32 | 45 46 47 48 | 61 62 63 64

BLOCKMASTER | BLOCKMASTER | BLOCKMASTER | BLOCKMASTER

THEORETICAL MASTER

(KBA)
KEY BITTING ARRAY

PAGE MASTER ______
GROUP MASTER ______
SECTION MASTER ______

PAGE # ______
OF ______
PAGES

SEQUENCE OF PROGRESSION

LINE MASTERS

RIZONTAL MASTER

VERTICAL MASTER	VERTICAL MASTER	VERTICAL MASTER	VERTICAL MASTER
1	17	33	49
2	18	34	50
3	19	35	51
4	20	36	52

ZONTAL MASTER

BLOCKMASTER	BLOCKMASTER	BLOCKMASTER	BLOCKMASTER
5	21	37	53
6	22	38	54
7	23	39	55
8	24	40	56

ZONTAL MASTER

BLOCKMASTER	BLOCKMASTER	BLOCKMASTER	BLOCKMASTER
9	25	41	57
10	26	42	58
11	27	43	59
12	28	44	60

RIZONTAL MASTER

BLOCKMASTER	BLOCKMASTER	BLOCKMASTER	BLOCKMASTER
13	29	45	61
14	30	46	62
15	31	47	63
16	32	48	64

BLOCKMASTER BLOCKMASTER BLOCKMASTER BLOCKMASTER

www.ingramcontent.com/pod-product-compliance
Ingram Content Group UK Ltd.
Pitfield, Milton Keynes, MK11 3LW, UK
UKHW051508290926
15471UKWH00008B/1503

9 781937 067090